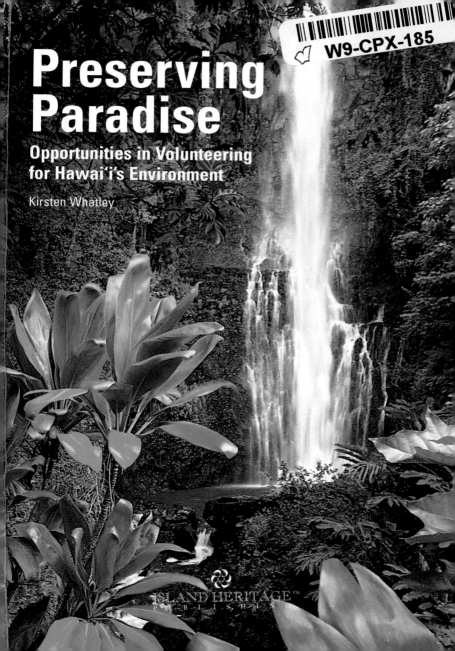

Preserving Paradise

Opportunities in Volunteering for Hawai'i's Environment

Kirsten Whatley

ISLAND HERITAGE™
PUBLISHING

ISLAND HERITAGE™
P U B L I S H I N G
A DIVISION OF THE MADDEN CORPORATION

94-411 Kō'aki Street
Waipahu, Hawai'i 96797-2806
For Orders: (800) 468-2800 • For Information: (808) 564-8800
Fax: (808) 564-8877
islandheritage.com

ISBN: 1-59700-578-9
First Edition, First Printing—2008

Cover Background Photo: Hanalei River, Kaua'i, by David Boynton.
Title Page Photo: Wailua Falls, Maui, by Ron Dahlquist.

ʻAʻohe hana nui ke alu ʻia.

No task is too big when done together by all.

—*ʻŌlelo Noʻeau: Hawaiian Proverbs & Poetical Sayings,*
Bishop Museum Press, 1983

Erica vonAllmen
Maui Restoration Group

Contents

Preface vii
Introduction viii

Multi-Island

Hawaiian Islands Humpback Whale National Marine Sanctuary 2
Hawaiian Islands National Wildlife Refuges 5
Hawai'i Nature Center 6
Hawai'i Wildlife Fund 7
Keep America Beautiful 8
Na Ala Hele 9
National Tropical Botanical Garden 10
The Outdoor Circle 12
Reef Check Hawai'i 14
Sierra Club—Hawai'i Chapter 17
     Hawai'i Service Trip Program 18
Surfrider Foundation 19

Hawai'i Island

Amy Greenwell Ethnobotanical Garden 21
Hakalau Forest National Wildlife Refuge 23
     Friends of Hakalau Forest National Wildlife Refuge 25
Hawai'i Hawksbill Turtle Recovery Project 26
Hawai'i Volcanoes National Park: Vegetation Program 28
Hawai'i Wildlife Center 30
Hawai'i Wildlife Fund—Hawai'i Island 33
The Kohala Center: ReefTeachers 35
Māla'ai: The Culinary Garden of Waimea Middle School 37
Mokupāpapa Discovery Center 39
Na Ala Hele—Hawai'i Island 41
Sierra Club—Moku Loa (Hawai'i Island) 43
Three Ring Ranch Exotic Animal Sanctuary 45
TREE (Tropical Reforestation and Ecosystems Education) Center Hawai'i 48

Every Day Is Earth Day 50

Kaua'i

Friends of Kamalani and Lydgate Park 53
Hanalei Watershed Hui 56
Hawaiian Monk Seal Conservation Hui 58
Hui o Laka: Kōke'e Natural History Museum 61
Kōke'e Resource Conservation Program 63
Sierra Club—Kaua'i 66
Surfrider Foundation—Kaua'i 68
Waipā Foundation 71

Maui & Moloka'i

Community Work Day Program 74
East Maui Animal Refuge (a.k.a. The Boo Boo Zoo) 76
Haleakalā National Park 79
    Friends of Haleakalā National Park 81
Hawai'i Nature Center—Maui 83
Hawai'i Wildlife Fund—Maui 85
Honokōhau Valley Project 88
Keālia Pond National Wildlife Refuge 90
Kīpahulu 'Ohana: Kapahu Living Farm 93
Maui Coastal Land Trust: Waihe'e Coastal Dunes and Wetlands Refuge 95
Maui Cultural Lands: Projects Malama Honokowai and Malama Hanaula 97
Maui Nui Botanical Gardens 99
Maui Restoration Group 101
Na Ala Hele—Maui 103
Native Hawaiian Plant Society 105
Pacific Whale Foundation 107
REEF (Reef Environmental Education Foundation) 109
    Project S.E.A.-Link 110
    FIN (Fish Identification Network) 111
Surfrider Foundation—Maui 112
Ka Honua Momona 114

Farm Apprenticeships

WWOOF 116

HOFA 118

O'ahu

'Ahahui Mālama i ka Lōkahi 120

Clean Water Honolulu 122

Friends of Hanauma Bay 124

Hawaii Audubon Society 126

Hawai'i Nature Center—O'ahu 128

Honolulu Zoo Society 130

Ho'oulu 'Āina: Kalihi Valley Nature Park 133

Lyon Arboretum 135

Mālama Na Honu 137

Mānoa Cliff Trail Project 140

Na Ala Hele—O'ahu 142

Nani 'O Wai'anae 144

O'ahu Invasive Species Committee 145

Sierra Club—O'ahu 147

Surfrider Foundation—O'ahu 149

Wild Dolphin Foundation 151

Acknowledgments 154

Appendix: Online Resources 155

Indexes

General 156

Project Subject 158

Time Commitment 162

Preface

Paradise is an ideal. It's somewhere other than where you spend your busy days. It's soaked in beauty, and the sight of it stops your mind. It lifts your heart.

But there's more beneath the surface of a pretty picture. There are struggles to keep Hawai'i's coastlines from being developed, desperate battles to protect endangered plants and wildlife against invasive species. And there are the tireless folks, who wake up obsessing about some cherished piece of wilderness or some underwater sanctuary or some species in danger of falling off the map.

You'll meet them here. Hopefully soon you'll be standing next to them.

Consider this an invitation, no matter your origin. Consider it a call to arms—and to hands and feet and backs and rakes and hoes and sweat. We all need a paradise to come home to. But we need to consider our impact on this paradise—more importantly, our responsibility to it. The opportunities in this book are one way to begin.

Mark Wasser/Wasser Photography

This māmane *flower and* koa *tree are two of the native species that programs such as the Sierra Club's Hawai'i Service Trip Program work hard to protect.*

Introduction

Why Volunteer?

How to describe that feeling that comes from giving? First, picture yourself on a lava rock outcrop, a mist of sea spray upon your face. You've just spent the night building cages in the sand, protecting sea turtle eggs from nonnative predators. Soon the hatchlings will have a chance to shuffle back to the ocean, swimming their way on to adulthood.

Or imagine you're thousands of feet high on an arid mountain trail. You retreat into the shade of a native *māmane* tree, a respite from the tropic sun, and marvel at your morning's work—a path once choked with alien weeds is now clear for other hikers to follow.

Maybe you're passionate about dolphins. Would a daily commitment to a research boat, recording spinner calf sightings, fit your idea of adventure with purpose? Or maybe you want to understand Hawaiian farming. You roll up your shorts and step into the thigh-high mud of a taro patch, helping out local farmers whose families have caretaken this land for generations.

This may not sound like the Hawai'i of the travel guides, and that's because it's not. The real Hawai'i can be hot and sometimes sticky. And there'll probably be bugs. Lots of bugs.

But talk with the volunteer from Minnesota, who just surfaced from a weekend in the Hakalau Forest National Wildlife Refuge, replanting native *koa* trees side by side with a local field biologist, restoring habitat for rare forest birds. She may not have a free mai tai glass to tuck into her suitcase. But the relationships she's built—with the land, its people, its culture— are lodged within her heart. Intangible. Meaningful.

Candace Calloway Whiting

Honu, *or Hawai'i's sea turtles, are a prime focus of many of the islands' marine protection efforts, such as Mālama Na Honu.*

One weekend of her time might not seem like much compared to all the work to be done. But she's one in a long chain of helping hands, each nurturing the *koa's* journey from seed to seedling, from potted plant to tree. One day, our volunteer will be standing beneath a whole grove of towering *koa* and know it might not have happened without her.

Whether you're visiting or you live in Hawai'i, whether you love snorkeling or hiking or native birds or sea turtles, there are passionate people doing just your kind of work. Give them a hand. It will give you a fish's-eye view of the islands. It will change your perspective of paradise.

So how to describe that feeling that comes from giving? You'll have to find out for yourself.

Restoring habitat for native birds such as the endemic short-eared Hawaiian owl, or pueo, is the passion of groups like the Hawaii Audubon Society.

About the Projects

There's a global trend happening called "voluntourism." In many cases, a placement agency handles the logistics of how you'll travel, where you'll stay, and what you'll do; in exchange, they charge you a deserved fee. The opportunities in this book are a little different.

They're more down-home, grassroots style—like calling up your auntie and asking if she needs help whacking the jungle away from her native plants, or an extra hand at the wildlife center where she volunteers, or helping protect the sea life that swims in the waters beyond her front door. You might be taken in as part of the family. There might be a potluck afterward. But how you get there and (with a few exceptions) where you stay are up to you.

Just as care is the only prerequisite for helping out your auntie, these opportunities cost nothing more than your time and the goodness of your heart. (Exceptions are organizations with membership requirements, and those who offer fee-based service trips along with their free activities. A few also request, but don't require, a donation.) Many groups featured in this book offer fascinating educational programs—some free and some with fees—and those are mentioned here too.

Remember, you don't have to take a year off of your life to make an impact volunteering. Many groups request only a few hours of your time, with none here requiring a commitment of more than three months. Some organizations prefer

Raising kalo *(taro), a traditional Hawaiian staple food, emphasizes the cultural as well as environmental focus of groups like Waipā Foundation.*

individuals, while others prefer groups, but everybody is welcome—visitor and resident alike. Ask Ed Lindsey, who's working to restore native forests and archaeological sites at Honokōwai Valley in West Maui: "It's important for visitors to feel a part of us, the Hawaiian community, to see Maui from the inside. Then when they go home, it's not just the sun and sand and sea they remember, but the people and our culture."

Culture and nature—as interdependent as a fragrant *maile* vine wrapped among the boughs of a rare *alani* tree. Restoring this relationship nourishes the spirit of all involved. So while the common thread among these opportunities is Hawai'i's environment—its diverse topography, its precious wildlife, its abundant sea—the projects here are just as much about people. They involve both the tireless folks mentioned earlier, who run the behind-the-scenes shows, and volunteers like yourself, who had a wild hair to drop what they were doing and join in for a larger cause, whether they're here just for a visit or have lived in Hawai'i their whole lives.

In the words of Moloka'i resident Todd Yamashita, who volunteers with the island's Ka Honua Momona on an ancient fishponds restoration project, "The most rewarding part is getting to know other people in the community.

Through meeting new people, you gain all kinds of different perspectives and insights into what it means to be Hawaiian and practice subsistence living."

How to Use This Book

While we've talked personally with each group featured in this book, these are not reviews. And we wouldn't want them to be—everyone's idea of preserving paradise is individual. What we have included are anecdotes from actual volunteers, to give you a window onto life in the field, firsthand. We hope this helps you make an informed decision of where to donate your time, what to help become a steward of.

The listings are ordered geographically by island. Since the reach of some groups spans more than one island, you'll find them in the Multi-Island section. Or maybe you don't care what island you end up on, you just know you want to work with wildlife, or on hiking trails, or with fishponds or anything having to do with native Hawaiian culture. Check the Project Subject index. If time is a concern, check the Time Commitment index.

While some groups request an advance notice of several months and a volunteer commitment of a couple more, most offer one-day affairs you can

© Krista Heide 2007

Organizations such as Reef Check Hawai'i survey the marine environment to better understand our human impact on its health.

This is the first volunteer resource of its kind for environmental opportunities in Hawai'i. We've made every effort to include all those groups who wanted their volunteer activities listed, though some may have gone unknowingly undetected. If you think your volunteer projects would be a good match for *Preserving Paradise*, please contact us with more information. We'll consider you for an upcoming edition.

And for those of you who volunteer with any of these groups, please share your experiences with us! We may choose your story to be included in the next *Preserving Paradise*.

Responses can be e-mailed to: preservingparadise@ welcometotheislands.com.

call about at the last minute. And this is important. Whether advance notice is required or not, *always call ahead* (or e-mail, if they prefer it). The information given here was up to date as of this book's writing, but things have a way of changing. Please contact your volunteer coordinators first—be sure you understand location and directions, times and duration, and what to bring. Since most activities are outdoors in the tropical elements, you'll also need protection like sunscreen and a hat, and often closed-toed shoes. Sometimes mosquito repellent. You should always carry water. And you might need to bring a lunch, as sites are often far from modern conveniences. Besides, you'll want to give your hosts a chance to prepare for your party—whether you're a party of one or ten.

Are you up for the adventure? Are you ready to give back to these islands that offer so much? Read on.

MULTI-ISLAND

Hawaiian Islands Humpback Whale National Marine Sanctuary

No matter how old you are or how many times you've seen them, you'll likely let out an incredulous shriek, you might even stand up and applaud the next time you see a whale leap out of the ocean. That's just the nature of these beautiful beasts. They literally take your breath away.

Warm Hawaiian waters are the birthplace of about two-thirds of the entire North Pacific population of endangered humpback whales. They return to Hawai'i from Alaska and other northern latitudes to breed, calve, and nurse their young. The migration takes about a month, one way, with the peak season for watching humpback whales being January through March.

The Hawaiian Islands Humpback Whale National Marine Sanctuary (NMS) is devoted to better understanding these mammoth sea mammals, while helping keep their underwater habitat healthy. Every winter, they run a Sanctuary Ocean Count, where volunteer whale watchers on Hawai'i Island, Kaua'i, and O'ahu simultaneously record the whales' behaviors—tail slaps, blows, full-body breaches, and dives. This valuable information then helps further our understanding of how humpback whales might be using nearshore waters.

It's a simple contribution that you can make onshore, from the comfort of your beach chair. What a way to enhance the world's knowledge of this little-understood creature.

NOAA HIHWNMS Photo

Volunteer Activities: Sanctuary Ocean Count, recording whale behavior.

When: Last Saturdays of January, February, and March, for 5 hours. (Preregistration is required—contact for upcoming session dates, requirements, site locations, and assignments.)

NOAA HIHWNMS *Photo*

Who: Both individuals and groups welcome. Children must be accompanied by a responsible adult, and require Sanctuary approval on a case-by-case basis.

Hardiness Level (5 being most difficult): 1 to 2 (Sites vary in accessibility, facilities, elevation, and exposure to elements. Ask about sites compatible with your physical condition.)

Advance Notice: Registration begins the previous December.

Education: See Web site for other public education opportunities throughout the islands, or visit the Sanctuary Education Center on Maui.

Donations: Monetary donations are appreciated and tax deductible if made to the National Marine Sanctuary Foundation. Contact regarding equipment and supplies needs.

Contact: See Web site for contacts on individual islands.
Hawaiian Islands Humpback Whale National Marine Sanctuary
6600 Kalaniana'ole Highway, Suite 301
Honolulu, HI 96825
For Hawai'i Island and O'ahu: (888) 55-WHALE, extension 253
For Kaua'i: (808) 246-2860
oceancount@noaa.gov
www.hawaiihumpbackwhale.noaa.gov

"My wife and I talked a lot about retirement. We knew we couldn't sit around in rocking chairs somewhere and grow old. I was lucky enough to get involved with [the Humpback Whale NMS] . . . it really is a continuation of my teaching work. And my wife was a librarian and is used to answering questions and being sure people get accurate information, so it's right in line. We see it as an opportunity to continue our education and learn more about the marine mammals.

"Most people we meet are very much like sponges—they're eager to absorb everything they possibly can. That eagerness to become informed and take that information back to their friends and family at home is rewarding. That's the part that keeps us going and makes it such a worthwhile experience.

"People kind of assume that we're 'experts,' that we know a lot about the humpbacks. And when we tell them there's a tremendous amount that nobody knows—that there's more we don't know than what we do—they find out they're actually getting in on the cutting edge of learning. They're being part of the discovery of the answers to the questions. It's something to add to the world's understanding of a little-known whale."

—Bruce Parsil
Kalāheo, Kaua'i

Hawaiian Islands National Wildlife Refuges

The National Wildlife Refuges (NWR) of the Hawaiian Islands are just that—places where birds can forage and nest in their native environment (the majority of Hawai'i's native land creatures are avian). They're places where wildlife can literally be at home, whether that's the forest, coast, or wetlands.

As development and alien species continue to threaten Hawai'i's native habitats, the U.S. Fish and Wildlife Service has designated ten such sanctuaries on the main Hawaiian Islands. They also manage remote refuges on a string of islands, reefs, and atolls about twelve hundred nautical miles to the northwest. Some of these remote regions are home to the world's most endangered duck, the Laysan teal.

U.S. Fish and Wildlife Photo

Many of the NWR sites are closed to the public, even volunteers, as the goal is to keep human impact at a minimum. But two refuges that do invite volunteer help are Hakalau Forest on the windward slopes of Hawai'i Island's Mauna Kea, and the wetlands of Maui's Keālia Pond. *(See listings in the Hawai'i Island and Maui sections for details.)*

Kirsten Whatley

Hawai'i Nature Center

Ants and millipedes, spooky spiders, even cockroaches become friends after a day at the Hawai'i Nature Center. It's two centers, in fact—an original field site in Makiki Valley on O'ahu and one on Maui, tucked into the foothills of magnificent 'Iao Valley. What started in 1981 as an offshoot of the Outdoor Circle (see listing later in this section) has become an outreach program for hundreds of thousands of Hawai'i's kids. (Oh, and adults are welcome too.)

Their mission: to help you fall in love with the outdoors. Their method: forest hikes, stream investigations, beach explorations, or just plain digging in the dirt. On Maui, you can also step into their Interactive Nature Museum, where you can get an up-close look at the wiggly world beneath the surface of 'Iao Stream, or high into the treetops of this storied valley.

Volunteering with the Hawai'i Nature Center might mean help with landscaping, removing invasive species, planting natives, taro patch restoration, or preserving the wetlands of a coastal marsh. *(See listings in the Maui and O'ahu sections for details.)*

Hawai'i Wildlife Fund

The Hawai'i Wildlife Fund (HWF) deals mainly with the fringe—the fragile ecosystem that rings the islands of Maui and Hawai'i Island, and the creatures that live there. Creatures like the endangered hawksbill sea turtles, which break away from familiar waters to make soft nests in the powdery sands of their birth. Or the Hawaiian monk seals, which regularly lumber their way onto shore to rest among sunbathers of a different species.

Started in 1996 by two former National Marine Fisheries Service scientists, Bill Gilmartin and Hannah Bernard, the HWF added Cheryl King to their team in 2000—first drawn in by a turtle nest watch on Lahaina Beach, Cheryl now passionately spearheads the hawksbill nesting monitoring program on Maui.

The devoted HWF crew also coordinates Makai Watch, where naturalists and volunteers patrol the shores, educating beachgoers about appropriate behavior around the coastline's archaeological sites, fragile reefs, and endangered marine species. HWF also helps keep the shorelines clean, dragging debris off the sand and out of the water, where it poses an ongoing threat to our aquatic neighbors.

Volunteering with the Hawai'i Wildlife Fund can mean jumping in on any one of these marine protection projects. (*See listings in the Hawai'i Island and Maui sections for details.*)

Michele Morris

Keep America Beautiful
Community Work Day Program and Nani ʻO Waiʻanae

An idea ahead of its time, Keep America Beautiful (KAB) has been influencing people to take responsibility for the environment since 1953. By uniting everyday citizens, businesses, and government, they've proved that together we can find solutions to reducing litter and waste—in essence, keeping our surroundings beautiful.

Nani ʻO Waiʻanae

Their network is vast: nearly one thousand affiliates and organizations span the fifty states. Every spring, their reach stretches even farther, when an estimated 2.8 million participants join the cause for the Great American Cleanup, the nation's largest community improvement effort. In 2007, over thirty thousand Great American Cleanup events took place. Volunteers collected two hundred million pounds of litter and debris, and planted 4.6 million trees, flowers, and bulbs. The numbers are almost too great to comprehend—a testament to the power of many hands.

There are KAB affiliates on each of the main Hawaiian Islands, with ongoing volunteer opportunities with Community Work Day Program on Maui, and Nani ʻO Waiʻanae on Oʻahu. *(See listings in the Maui and Oʻahu sections for details.)*

Na Ala Hele

One definition of the words *ala hele* is the "way to go." In 1988, the state Department of Land and Natural Resources decided that preserving historic trails and ensuring access to the wild was definitely the way to go. They established the Na Ala Hele Trail and Access System on every island, so that people could keep on hiking, camping, fishing, hunting, picnicking, and photographing in Hawai'i's wilderness. Of course, conservation goes both ways—maintaining the land for nature enthusiasts also prevents the wildlife habitat from succumbing to development.

Along with keeping the current trails cleared and stabilized, there's often trailblazing of new footpaths needed. As Virginia Aragon, volunteer coordinator for Na Ala Hele on Hawai'i Island, describes: "When I first walked the rough unfinished trail of Kīpuka 21 [a Hawai'i Island rain forest], I tried to envision its completion and what major work it would take to get there. I was a bit intimidated by what appeared to be a daunting task. But just as fast as that thought crept into my mind, it was soon kicked out by the overwhelming response from the community. Volunteers heeded the *kīpuka's* call for help and responded by giving all they had. . . . [It has] helped move us just that much closer to our goal: the completion of trail built in an ancient pocket of forest, a sanctuary for native birds and native plants, intended for the enjoyment and education of the people and visitors of Hawai'i." (*See listings in the Hawai'i Island, Maui, and O'ahu sections for details.*)

Greg Lindstedt

Lea Taddonio

National Tropical Botanical Garden

The warm, wet belt that encircles the planet, the part that spins closest to the sun, is what we know as the tropics, that area just above and below the equator. It's home to a remarkable 90 percent of the planet's plants and animals—home also to the highest extinction rate in the world.

The National Tropical Botanical Garden (NTBG) works to intercept this unfortunate statistic. Through research and education, propagation and restoration, the gardens serve as botanical arks for at-risk tropical species from around the world. Four of the NTBG gardens are in Hawai'i—three on Kaua'i and one in Hāna, Maui. A fifth garden is in Florida. As the only tropical climate zones in the United States, these locales are ideal homes away from home for the wild-collected species. Equally important is the gardens' emphasis on native Hawaiian plant conservation, habitat restoration, and the perpetuation of traditional local knowledge.

The gardens range from steep, jungled mountainside, to a tapestry of microclimates in a single valley, to the site of Hawai'i's largest *heiau* (ancient place of worship), an impressive lava-rock structure believed to date back to the thirteenth century. Volunteers can take part in the ongoing work of horticulture and conservation at these sites, or can lend a hand in the NTBG nursery. *(Contact the main headquarters for opportunities on both islands.)*

Volunteer Activities: Horticulture, conservation, nursery work.

When: Year round, flexible hours.

Who: Both individuals and groups welcome. Young children supervised by their parents accepted on a case-by-case basis.

Hardiness Level (5 being most difficult): 2 to 3

Advance Notice: 1 week preferred.

Education: Garden Connections is a series of free public classes in horticulture, art, culture, and health. (Check the News and Events section on their Web site for upcoming classes.)

Donations: Monetary donations are appreciated and tax deductible. Contact regarding equipment and supplies needs.

Contact: E-mail preferred.
Lea Taddonio
National Tropical Botanical Garden
3530 Papalina Road
Kalāheo, HI 96741
(808) 332-7324
volunteer@ntbg.org
www.ntbg.org

"I do not have a green thumb. In fact, I would say I have a black thumb! However, the NTBG nursery staff were so patient and friendly that soon I was potting rare and native plants like a pro. It was a wonderful and relaxing morning on my vacation."
—Donna Stengel
Manchester, Michigan

"I got to work in the Canoe Garden, if you want to call it that. I called it fun. I learned about the plants ancient Polynesians brought with them on their overseas voyages. I also got to visit a waterfall that's not included on any tour."
—Dave Holmes
Melbourne, Australia

Lea Taddonio

The Outdoor Circle

Clean, green, and beautiful: a simple vision for Hawai'i's environment. Since 1912, the Outdoor Circle has been living by this motto, planting trees, protecting green spaces, and perhaps most notably, preventing billboards from populating the islands' magnificent landscape—a landmark decision upheld since 1926.

With thirteen branches scattered across the main Hawaiian Islands, there's always an Outdoor Circle nearby, helping local communities keep their neighborhoods green. The most active in volunteer opportunities are the Kona, North Shore, Lani-Kailua, and Waimea branches.

In Kona, on Hawai'i Island's western coast, volunteering might mean helping on a landscaping project, or with outreach and event staffing for a fund-raiser, or maybe on a field trip. On O'ahu's North Shore, Outdoor Circle volunteers are often found maintaining the bicycle path that stretches 3 1/2 miles along dreamy Sunset Beach. Other times, they're out sprucing up public spaces, like the lovingly landscaped rotary circle in Hale'iwa.

At Hawai'i Island's Lani-Kailua branch, Outdoor Circle volunteers might find themselves beautifying parks, schools, or roadsides by planting and mulching shade trees, cleaning up unlawful signs and banners, or planning for the next stage of beautification along Kailua Road.

The Waimea Outdoor Circle on Hawai'i Island puts most of its beautification efforts into Ulu Lā'au ("garden of trees"), also known as Waimea Nature Park. This rich concentration of native plantings spans ten acres in the heart of Waimea town. There is an ongoing need to clear invasive species, reintroduce native plantings, weed, mulch, prune,

"For the last twelve or thirteen years, I've been in charge of the bike path beautification. It's there because it used to be a railroad. We've modeled it after the nationwide Rails to Trails program—across the country they're turning railroads into bike trails.

"It's been fabulous. I've made great, great friends, a lot of older women my mother's and grandmother's age, and I really enjoy those connections. You're socializing as well as doing good work. And as a teacher, I'm modeling the behavior I'd like to see in my students, and my community.

"To see children and moms rolling strollers down the bike path, and elders taking walks . . . having a safe walkway people can get from one place to another on is the greatest benefit. It's really convenient, it's safe, and it's beautiful."

—Rex Dubiel-Midkiff
Sunset Beach, O'ahu

and much more. *(See the Outdoor Circle Web site or contact the main office for details and opportunities on both islands.)*

Volunteer Activities: Landscaping, maintaining public spaces, fund-raising.

When: Year round. (See Web site for each branch's schedule.)

Who: Individuals preferred; contact regarding groups. All ages welcome.

Hardiness Level (5 being most difficult): 1 to 4

Advance Notice: 2 weeks.

Education: See Web site for each branch's schedule.

Donations: Monetary donations are appreciated and tax deductible. (You may designate a specific branch for the donation's use.)

Contact: E-mail preferred.
The Outdoor Circle
1314 South King Street, Suite 306
Honolulu, HI 96814
(808) 593-0300
mail@outdoorcircle.org
www.outdoorcircle.org

Roger H. Williams

© Krista Heide 2007

Reef Check Hawai'i

Never underestimate the power of a small idea. Or the large effect of a small idea once multiplied across the globe.

The first Reef Check was held on Kaua'i in 1997, drawing two hundred volunteers to help assess the island's offshore topography and human effects on it. Volunteers looked for fish like *uhu* (parrot fish), indicators of a reef's health, and *ula* (spiny lobster), whose absence points to overfishing. They sized up the ocean floor and the coral itself, looking for symptoms of disease and damage.

By year's end, Reef Check had conducted surveys in thirty other countries around the planet. Using methodology standardized by UCLA marine ecologist Gregor Hodgson, this broad snapshot encompassed three hundred reefs throughout the Atlantic, Caribbean, and Indo-Pacific oceans. The results proved that, indeed, our reefs were in dire need of attention.

Today, Reef Check operates in over eighty countries and has enlisted an army of ocean-loving volunteers. The data they collect helps marine scientists keep tabs on these rain forests of the sea. To join the Reef Check Hawai'i family, you simply have to be a snorkeler or scuba diver. One-time trainings in reef ecology and survey methods are then followed by in-water training surveys, after which you're qualified to join Reef Check teams throughout the main Hawaiian Islands.

Reef Check Hawai'i also hosts quarterly beach cleanups on these islands, expanding their practice of monitoring underwater health to ensuring that adjacent shorelines also stay pristine.

"I volunteer with Reef Check so that I can learn more about life below the water while assisting with making life better above the water! This is a fun way to not only snorkel, but to possibly make a difference in keeping Hawai'i's reefs healthier. Healthy reefs mean healthier islands."

—Anna Myers
Princeville, Kaua'i

© *Reef Check Hawai'i*

"Reef Check Hawai'i has an amazing marine ecology course. In two days of training I learned as much as it takes a whole semester in school to learn about the marine ecosystem. Although our actual dive to perform the reef check was cancelled due to high surf, Krista and Stephanie scheduled a new training dive for us, and Stephanie was so dedicated that she brought us out on the reef check while she was eight months pregnant. She was more at home swimming in the water than walking on land. It was amazing to see the dedication and love they have for the ocean and its creatures.

"For years I have been hearing many things about alien *limu* [seaweed] and what it can do to harm the coral reefs, however no one could ever show me what the *limu* looked like. Reef Check [distributes] a small waterproof pocket guide [produced by the Hawaiian Division of Aquatic Resources] that shows many of the alien *limu*, what they look like and how to protect from spreading it. This impressed me to see how organized and constructive they are in producing new tools and innovative ideas to help save our coral reefs. I am very proud to be a Reef Check volunteer."

—Scott Bacon
Kīlauea, Kaua'i

Volunteer Activities: Reef Check surveys and beach cleanups.

When: Reef Check surveys—at least 2 sites on each main island per month, for 2 to 3 hours. Beach cleanups—every quarter on each main island, for 2 to 3 hours. (Free volunteer training required before surveys— Introduction to Coral Reef Ecology and Threats to the Reef, 2 hours; Introduction to Reef Check Survey Methodology and Indicator Species Identification, 2 hours.)

Who: Both individuals and groups welcome. Kids must be 15 or older to participate in in-water reef surveys.

Hardiness Level (5 being most difficult): 1 to 5, depending on personal experience.

Advance Notice: 1 week preferred.

Donations: Monetary donations are appreciated and tax deductible. Contact regarding equipment and supplies needs.

Contact: E-mail preferred.
Krista Heide
Reef Check Hawai'i
P.O. Box 621
4224 Wai'alae Avenue
Honolulu, HI 96816
(808) 953-4044
contact@reefcheckhawaii.org
www.reefcheckhawaii.org

Sierra Club—Hawai'i Chapter

If you want to go way back, the Sierra Club started in 1892, when California naturalist John Muir and his cohorts formed the club to, in his words, "make the mountains glad." Since then, his successors in Hawai'i have been making the islands glad by fighting legislative battles to protect this fragile ecosystem, leading folks on wilderness outings, and running a successful volunteer program.

With local groups on all major Hawaiian Islands, the Sierra Club coordinates volunteer projects throughout the state. Join them to help remove invasive plants, build fences to keep feral animals at bay, and maintain miles of trails that wind through the islands' wilderness. *(See listings in the Hawai'i Island, Kaua'i, and O'ahu sections for details.)*

"Working on a service trip to help preserve Hawai'i's native dryland forests was so rewarding. Many of these ecosystems are on the brink of extinction and it's great to feel that even small actions, such as removing invasive weeds or outplanting endangered plants, can be significant. Working with HSTP allowed me to go to an area of Hawai'i I probably would not have seen otherwise, and the camaraderie on the trip was great."

—Sarah Trask
Honolulu, O'ahu

Jamie Tanino

Hawai'i Service Trip Program

An extension of the previous listing is the Sierra Club's statewide Hawai'i Service Trip Program—one- to two-week adventures designed to give you an in-depth experience of the wild while you do hands-on work to help maintain it. The service trip fee is $150 ($100 for students), including accommodations, meals, and transportation; Sierra Club membership is required.

Recent trips took volunteers to Pu'u Wa'awa'a on Hawai'i Island—once home to the most diverse dry and semidry forests on all the island, it now battles invasion from nonnative plants and animals. Volunteers bunk in a wilderness cabin at 4,100 feet and spend their days planting native vegetation and repairing fence line, while leaving time to explore the nearby forest bird sanctuary.

The Hawai'i Service Trip Program offers projects on Hawai'i Island, Kaua'i, Lāna'i, Maui, Moloka'i, and O'ahu. Locations are extremely beautiful and often remote, so a love for being "out there" is essential.

Volunteer Activities: Service trips throughout the islands.

When: Several times per year, for 1 to 2 weeks. (Lodging available in volunteer cabins.)

Who: Both individuals and groups welcome, age 16 and older; guardian signature required for those under 18. Maximum participants per trip limited to 12. Locations are often far from medical care—participants must be in good health.

Hardiness Level (5 being most difficult): 3 to 5

Advance Notice: 2 months preferred.

Donations: Monetary donations are appreciated and tax deductible. Contact regarding equipment and supplies needs.

Contact: E-mail or postal mail. (See Web site for current e-mail contacts.)
Hawai'i Service Trip Program
P.O. Box 2577
Honolulu, HI 96803
www.hi.sierraclub.org/hstp

Surfrider Foundation

In 1984, a handful of Malibu, California, surfers had the vision to band together and protect the places they loved most—the beaches, the waves, the coastal ecosystems, and public access to it all. They formed Surfrider Foundation, which now has over fifty thousand members in eighty chapters around the world.

Their members aren't only surfers, they're also those who love to ride waves with their bodies, bodyboards, sailboards, and kiteboards. Some call themselves marine conservationists. Some simply love the sea.

In Hawai'i, Surfrider Foundation keeps busy cleaning up local beaches, addressing water quality and polluting runoff, speaking out against coastal overdevelopment, restoring native plantings in beach parks, and teaching others how to surf. (*See listings in the Kaua'i, Maui, and O'ahu sections for details.*)

Marvin Heskett

HAWAI'I ISLAND

Peter S. Goltra for the National Tropical Botanical Garden

Noa Lincoln

Amy Greenwell Ethnobotanical Garden

What esteemed botanist Amy Greenwell (1920–1974) used to call "pre-Cookian" referred to the plants growing in her garden. Endemic, indigenous, and Polynesian-introduced species, they defined the Kona region before the arrival of Captain Cook, the first recorded European to set foot in Hawai'i.

In a town now named Captain Cook, her garden still grows. Lovingly tended to by Bishop Museum staff, over two hundred species fill this fifteen-acre Eden. A true Kona *ahupua'a* (ancient land division), its planting zones range from upland forest to the coast.

The garden's focus is on ethnobotany—the relationship between people and their plants, whether for food, medicine, shelter, transportation, or culture. Many plants are, in fact, growing within ancient archaeological sites on the property. Also featured is an insect house that harbors Kamehameha butterflies, *pulelehua*, one of only two species of butterfly native to the Hawaiian Islands.

"My first time in Hawai'i was in 1996—we came to the gardens and walked the grounds. It was just a wonderful place to be. It has wonderful *aloha* spirit. The native plants were also a major attraction, because you don't often get to see them [in the wild].

"[Volunteering] is all that I could have hoped for. The staff is very friendly and supportive. They have great relations with the community, many people who've been here for generations and who are very knowledgeable about plants and the [Hawaiian] culture.

"The garden is a repository for knowledge that's in danger of being lost—not only of endangered plant species, but of the culture as well. It's a safe haven in many ways."

—Priscilla Studholme
Kailua-Kona,
Hawai'i Island

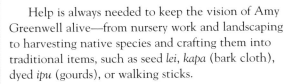

Help is always needed to keep the vision of Amy Greenwell alive—from nursery work and landscaping to harvesting native species and crafting them into traditional items, such as seed *lei*, *kapa* (bark cloth), dyed *ipu* (gourds), or walking sticks.

Volunteer Activities: Nursery work, making crafts, landscaping.

When: Year round, from 1 hour to 1 day per project.

Who: Both individuals and groups welcome. Age minimums vary—contact for details.

Hardiness Level (5 being most difficult): Nursery work, making crafts—1 to 2. Landscaping—2 to 5.

Advance Notice: Preferred—contact for details.

Education: Public garden tours, workshops, and service learning trips are available with 1-day advance notice. (See Web site for details and fees.) Free kids' programs include a Propagation Club for younger kids, and paid internships for teens and preteens. (Contact for more details.)

Donations: Monetary donations are appreciated and tax deductible. Contact regarding equipment and supplies needs. Ask about the Adopt-a-Tree program.

Contact: E-mail or phone.
Noa Lincoln
Amy B. H. Greenwell Ethnobotanical Garden
P.O. Box 1053
Captain Cook, HI 96704
(808) 323-3318
agg@bishopmuseum.org
www.bishopmuseum.org/greenwell

"We've been going pretty much every Wednesday for a couple years. I really like the hands-on experience, and all the connections you make and the friends you meet. You get to work with native Hawaiian plants, and you work with professionals.

"When I first went up there, I knew almost nothing about native Hawaiian plants. Now I know scientific names, botanical names, lots of things that will be useful for later studies."

—Matt Sylva
Student
Kailua-Kona,
Hawai'i Island

Hakalau Forest National Wildlife Refuge

Those who've been to Hakalau pepper their description with words like "spiritual," "magical." Home to fourteen native bird species (eight of them endangered), twenty-nine rare or endangered plants, and Hawai'i's only native land mammal, the 'ōpe'ape'a (Hawaiian hoary bat), the region is definitely a rarity.

Nearly thirty-three thousand acres comprise the refuge. On terrain lower than four thousand feet, it's a land rich in rain—about 250 inches fall annually; farther upslope, that can drop to around 100 inches. Along the way, bogs and forests give way to cattle pasture dominated by invasive grasses and weeds.

Volunteers are helping eradicate the alien species; they're collecting native seeds, propagating them in the greenhouse, and later replanting them back in nature. As volunteer Jackson Bauer puts it: "It's exciting to envision that one day in the future, I can bring my grandchildren here and watch the rare forest birds thrive in the shade of the koa trees that I helped plant." On other occasions, volunteers may be asked to help the refuge staff band nēnē, assisting in the research and protection of the endangered Hawaiian goose.

Hakalau is a refuge in many ways, and its volunteer slots fill up about a year in advance. You can also check the Friends of Hakalau Forest National Wildlife Refuge (NWR) listing that follows for additional opportunities.

(See Multi-Island section, under Hawaiian Islands National Wildlife Refuges, for more about the agency itself.)

Volunteer Activities: Greenhouse work, seed collecting, weed control, tree planting.

When: Year round, for 1 to 3 days.

Who: Groups preferred, but individuals accepted if committed for several weeks or months, age 15 and older. Volunteers must be physically fit and able to work at an elevation of 6,500 feet (the refuge is 2 hours from the nearest medical facilities).

Hardiness Level (5 being most difficult): 2 to 4

Advance Notice: 1 year.

Education: Public access to the refuge is granted only at Maulua on weekends and holidays, by reservation (contact the office).

Donations: Monetary donations accepted through Friends of Hakalau Forest NWR (see listing that follows). Contact the refuge regarding equipment and supplies needs.

Contact: Postal mail preferred.
Refuge Manager
Hakalau Forest National Wildlife Refuge
60 Nowelo Street, Suite 100
Hilo, HI 96720
(808) 443-2300
www.fws.gov/pacificislands/wnwr/bhakalaunwr.html

Friends of Hakalau Forest National Wildlife Refuge

Spreading the good word of what's happening at the refuge keeps the Friends of Hakalau Forest NWR quite busy. They're the nonprofit arm set up to accept donations for the refuge during its restoration and maintenance. (All donations are tax deductible.)

Then there are the much-anticipated all-day work parties, when they bring volunteers out to Hakalau—their schedule is variable, but the projects require less advance notice than those of the NWR agency itself. Contact them for upcoming dates. (You must be over eighteen and in good physical condition.)

friendsofhakalauforest@gmail.com
www.friendsofhakalauforest.org

"The volunteer program at Hakalau Forest is truly a unique and exceptional experience! The belief that this place has a spirit and beauty that needs care and commitment comes through in the ways of the refuge staff, the caretakers. Working with them on the refuge I felt invigorated and instilled with an increased desire to commit more time to volunteering for the sake of our *ʻāina* [land] and her creatures."

—Carissa Dwelli
Hilo, Hawaiʻi Island

U.S. Fish and Wildlife Photo

Hawaiʻi Hawksbill Turtle Recovery Project

The endangered *honu ʻea* (hawksbill turtle) nests on some of Hawaiʻi Island's most rugged and remote beaches—less than one hundred nesting turtles have been documented in the Hawaiian Islands, with over 90 percent nesting here. On average, a female lays more than 175 eggs at a time and can lay up to six nests per season. After nesting, she then lumbers back to sea, leaving the eggs to incubate beneath the sand for the next two months.

Then the babies hatch. And in a remarkable feat of teamwork, they move the sand in their nest from ceiling to floor, until they've raised themselves to just beneath the surface of the beach. When conditions are right, they break through the sand and scramble toward the ocean in a mad collective dash.

But the process is perilous. Feral rats, cats, and mongeese prey on the nests and tiny turtles. The hatchlings can get caught in rocks or human footprints, tangled in the shoreline vegetation, or confused by artificial lights, which may draw them away from the water and leave them stranded to die.

Volunteers are the true champions of the Hawaiʻi Hawksbill Turtle Recovery Project. Hiking up to ten miles across raw terrain, carrying a heavy backpack, and camping for six-plus consecutive nights is a commendable feat. Turtle-monitoring shifts run from dusk into the wee hours, during which volunteers watch for signs of turtle activity, tag nesting turtles, build protective cages over the nests, help hatchlings reach the ocean, and trap predators. Not only is this a grand contribution to the protection of an endangered species, the experience is unforgettable.

Volunteer Activities: Hawksbill turtle monitoring. *(Euthanization of nonnative predators— rats, mongeese, and feral cats—is required.)*

When: May to December, 10-week minimum commitment. (Lodging available in volunteer housing and at campsites. Daily meal stipend provided.)

Who: Individuals preferred, age 18 and older. Volunteers must have a driver's license.

Hardiness Level (5 being most difficult): 5

Advance Notice: 2 months or more preferred.

Donations: Monetary donations are appreciated (not tax deductible).

Contact: E-mail preferred.
Will Seitz
Hawai'i Hawksbill Turtle Recovery Project
P.O. Box 52
Hawai'i Volcanoes National Park, HI 96718
HAVO_turtle_project@nps.gov
www.nps.gov/havo/naturescience/turtles.htm

"It involves a lot of hiking and being on beautiful beaches in Hawai'i. We hike an average of eight miles a day over lava rocks. It's pretty intense. We'll camp at one beach, then we'll check other beaches all along the coastline.

"It's a people project as much as a turtle project. I've learned so much about spending time with other people, and from other cultures—because you spend a week at a time with just one or two other people, and we have volunteers from all over the world.

"And I've learned about the culture of Hawai'i—I've been here on vacation, but living here is a whole different thing."

—Anne Farahi
Elizabeth, New Jersey

Hawai'i Hawksbill Turtle Recovery Project

Hawai'i Volcanoes National Park
Vegetation Program

Say "Hawai'i Island" and most people think of lava—either bursting from the cauldron of Kīlauea Volcano or carving its searing path toward the sea. But Hawai'i Volcanoes National Park has other treasures—like twenty-three species of endangered plants, including fifteen tree species, and six of Hawai'i's fifteen endangered native birds.

These species have survived here for millennia. Yet in the last two hundred years, a foreign invasion of plants, animals, and insects has begun to push them toward extinction. In a habitat as diverse as Hawai'i Volcanoes National Park, with seven ecological life zones ranging from alpine to seacoast, saving these original inhabitants is essential.

The park's Vegetation Program takes volunteers into these delicate ecosystems to help remove alien plants, plant native species, and collect seeds. Groups are welcome for day trips; individuals are accepted with a commitment of three months or more. This longer-term work is for the more adventurous, involving camping in all kinds of weather in the striking terrain of the backcountry.

NPS Staff

Volunteer Activities: Alien plant control, planting of native seedlings, seed collection.

When: Year round. (Lodging and daily meal stipend provided to long-term volunteers.)

Who: Groups welcome for day projects; individuals accepted with a commitment of three months or longer. (Individuals must have a driver's license.) Kids must be supervised by an adult.

Hardiness Level (5 being most difficult): 3 to 5

Advance Notice: 2 months preferred.

Education: Ranger-led public interpretation programs, Junior Ranger programs, and special speaker presentations are arranged through the park visitors' center. (See Web site for more information.)

Donations: Monetary donations are appreciated and tax deductible. Contact regarding equipment and supplies needs.

Contact: E-mail preferred. For groups—Susan Dale, Susan_Dale@contractor.nps.gov, (808) 985-6195. For individuals—Corie Yanger, Corie_Yanger@partner.nps.gov, (808) 985-6097.
Hawai'i Volcanoes National Park
Vegetation Program, Division of
    Resources Management
P.O. Box 52
Hawai'i Volcanoes National Park, HI 96718
www.nps.gov/havo/

"It is exhilarating to be in areas not often seen by the public. The native birds are singing in the trees while we plant, collect seed, or monitor an area. One can see the mist on the trees in the morning and later seek the shade of a tall native tree as the day warms.

"It is very rewarding to return to areas where we have been working and see the regeneration processes taking place in the native forest. Changes can be quite rapid and dramatic.

"In addition . . . we have found a wonderful community of people. All of the people who work and volunteer in the park are interested in the same cause and giving back to the community. They are very inclusive, and even as volunteers we feel like full members of the team and the appreciation of our contributions."

—Mark and
Carol Johnson
Hilo, Hawai'i Island

© Robert Shallenberger

Hawai'i Wildlife Center

Billions of gallons of petroleum ship into Hawai'i's ports each year, making oil spills infrequent but inevitable. And it does happen. In 1996, thirty-nine thousand gallons spilled into Waiau Stream and Marsh, and ultimately Pearl Harbor. In 2001, a Japanese ship carrying sixty-five thousand gallons of oil collided with a navy submarine nine miles off O'ahu's southern coast.

When oil coats a seabird's feathers, they lose their waterproofing and the down beneath the feathers gets wet. The bird then loses its buoyancy and becomes hypothermic.

The Hawai'i Wildlife Center plans to cater to such disasters. In fact, they'll be organized to accept any form of sick or injured native wildlife— seven days a week, year round. After treatment, they'll rehabilitate the creature and release it back to its natural home.

Linda Elliott is the inspiration behind the new center, which, in addition to the wildlife response unit, will feature an education pavilion and

"I got a grant from UH Hilo, and we decided to do a *nēnē* education project with our [Waimea Middle School] students. Linda really helped us, and we worked with the Big Island Country Club educating golfers, so they understand *nēnē* behavior and that *nēnē* are special, and our state bird.

"It's really exciting for the kids to see the real-world component of what they're studying in class. Linda has a lot of respect for wildlife, and it was great for the kids to be around that kind of energy."

—Lily Edmon
Waimea, Hawai'i Island

"I got the grant with Lily Edmon to do a *nēnē* education program with the kids and Big Island Country Club. Linda has had so much experience in animal rehabilitation. And she's an amazing networker and resource for information about the *nēnē*.

"There's been a human-*nēnē* conflict at the country club, with *nēnē* nesting on the golf course. The kids did presentations and PowerPoints to the golfers and the public, telling why the *nēnē* are special and sacred to the islands.

"The kids were so excited, working with adults in the profession, and seeing the receptiveness of the golfers and the community. Being able to participate in these projects opened my mind to what was possible in making a curriculum.

"The most rewarding part for me was seeing what role the kids can have, as ambassadors, within their community."

—Jessica Schwarz
Boulder, Colorado

interpretive courtyard. Part of an international team of wildlife rescuers, Linda has put her skills to practice in South Africa, Spain, the Galápagos Islands, and here in Hawai'i. She's raised two-thirds of the money needed to complete the center, to be constructed on two acres of donated land in Kohala. In the meantime, Linda has been running youth programs and giving public presentations, educating others on how to protect Hawai'i's native wildlife, such as the endangered state bird, the *nēnē* (Hawaiian goose).

Volunteers can help bring the Hawai'i Wildlife Center to life by offering administrative or fund-raising support, helping with construction, or landscaping the center's grounds. As Linda puts it, "Hawai'i has the rarer species in the United States, and protecting them is an exciting opportunity. This is a chance to be involved from the ground up."

Volunteer Activities: Administrative support, fund-raising, construction, landscaping.

When: Year round, flexible hours.

Who: Individuals preferred, age 18 and older.

Hardiness Level (5 being most difficult): 1 to 5

Advance Notice: 1 month.

Education: Free environmental education programs (on Hawai'i's native plants, animals, and ecosystem) can be arranged for groups or individuals, with 1-month advance notice.

Donations: Monetary donations are appreciated and tax deductible. Contact regarding equipment and supplies needs.

Contact: E-mail preferred.
Linda Elliott
Hawai'i Wildlife Center
P.O. Box 551752
Kapa'au, HI 96755
(808) 889-5180
info@hawaiiwildlifecenter.org
www.hawaiiwildlifecenter.org

© Robert Shallenberger

Hawai'i Wildlife Fund—Hawai'i Island

The southern coastline of Hawai'i Island is home to swift currents and lashing waves. It's also a natural repository for tons of marine debris—literally. Ten to twenty tons each year. And the trash keeps coming. Thousand-pound nets, tangled fishing lines, all sorts of plastic, even tires wash up on shore, where endangered monk seals and hawksbill turtles come to rest and nest. Offshore, humpback whales, dolphins, seabirds, and numerous fish are repeatedly entangled in the debris or fatally ingest it. Then there's the damage done when the wind and ocean currents drag it mercilessly across the coral reef.

In 2003, the Hawai'i Wildlife Fund (HWF) became a steward of this threatened landscape. Since then, HWF and community volunteers have hauled away over ninety tons of marine debris. Some of it goes to the landfill, while the fishing net and line are sent to Honolulu, where they're transformed into usable electricity at a trash-to-energy conversion plant.

On select Saturdays you can find dedicated HWF staff and volunteers continuing this effort somewhere between the Wai'ōhinu ("shining water"), or Kamilo, coast and Ka Lae, or South Point—about nine miles of coastline. They'd happily welcome your helping hands.

(See Multi-Island section, under Hawai'i Wildlife Fund, for more about the organization itself.)

Volunteer Activities: Beach cleanups.

When: Select Saturdays, about every 2 months, for 5 hours. (Contact for upcoming dates.)

Who: Both individuals and groups welcome. Small children must be supervised by an adult. (4x4, high-clearance vehicles are required to reach most sites.)

Hardiness Level (5 being most difficult): 4

Advance Notice: None needed.

Donations: Monetary donations are appreciated and tax deductible. Contact regarding equipment and supplies needs. Ask about the Adopt-a-Whale/Dolphin/Monk Seal/Turtle/Coral Reef programs.

Contact: E-mail or phone.
Bill Gilmartin
Hawai'i Wildlife Fund
P.O. Box 70
Volcano, HI 96785
(808) 985-7041
bill-gilmartin@hawaii.rr.com
www.wildhawaii.org

All photos by Bill Gilmartin

The Kohala Center
ReefTeachers

Kahalu'u Bay, Kona. It's one of the most popular snorkeling spots in Hawai'i, host to four hundred thousand visitors a year. Some have never snorkeled before. Many have never had the need to distinguish live coral from dead rock or are aware that coral is a living organism. As a result, the underwater landscape of Kahalu'u is in danger.

That's what prompted the University of Hawai'i Sea Grant to develop ReefTeach—a volunteer program facilitated by the Kohala Center to help people protect Hawai'i's coral reefs. Using picture boards, reference materials, and identification books, the message is "low impact" and is always shared with *aloha*. It may also include how to respect a basking turtle or how to interact with the fish and other wildlife at the bay.

Adults and teens are welcome to join the ReefTeach team after a one-time on-site training session of about two hours. (Shorter trainings may be available for volunteers focusing on a single topic, such as sea turtles.) The teaching then takes place onshore at this spectacular beach.

Volunteer Activities: Public outreach as a ReefTeacher.

"My greatest pleasure is sharing my knowledge of the reef with the children. It never ceases to amaze me how many will stop by our booth and listen to what we have to say about the ecology of the reef. They then go on their way and after snorkeling will come back and share with us what they have seen, oftentimes spending hours going through our books on fish and talking story with us.

"If we can reach one of them we have done our task."

—Jim Bausano
Kailua-Kona,
Hawai'i Island

R. Magnus

"One day there was a turtle that was grazing on the algae-covered rocks in the shallow water. A small group of children were listening carefully to what I was telling them about the turtle. A little later a mom was wanting to get a photo of her daughter with the turtle, and she told [her] to get a little closer. She replied, 'No, Mom, I can't go closer than ten feet or I will disturb him. A photo isn't as important as that.' She then repeated to her mom many of the things I had said.

"[Her mom] came up and thanked me afterward, and said her daughter thought that this experience was the best of all her vacation."

—Christine Sheppard
Kailua-Kona,
Hawai'i Island

When: Year round, flexible days, for 3-hour shifts. (Free volunteer training required—1 hour of presentation; 1 hour in the water. Shorter trainings available for volunteers focusing on a single topic.)

Who: Both individuals and groups welcome, age 12 and older. Volunteers must be able to swim.

Hardiness Level (5 being most difficult): 1

Advance Notice: 2 to 3 weeks.

Education: Public education is free—contact for details.

Donations: Monetary donations are appreciated and tax deductible. Contact regarding equipment and supplies needs. Ask about Adopt a Day at Kahalu'u Bay.

Contact: E-mail preferred.
Cindi Punihaole
The Kohala Center
P.O. Box 437462
Kamuela, HI 96743
(808) 887-6411 or (808) 895-1010
cpunihaole@kohalacenter.org
www.kohalacenter.org

R. Magnus

Māla'ai
The Culinary Garden of Waimea Middle School

Before digging a new garden bed or turning the compost, before transplanting lettuce seedlings or weeding around the sweet potato patch, there must first be two minutes of silence. The kids sit on the earth, spread throughout the

Amanda Rieux

garden, and are simply quiet, observant. When garden class starts, the change in relationship between the students and their food has already begun.

This is the aim of Māla'ai, the culinary garden at Waimea Middle School: to teach kids not just how to organically grow, harvest, and prepare their food, but in what spirit it should be shared. Led by garden savant Amanda Rieux, the program is modeled after Alice Waters's successful Edible Schoolyard in Berkeley, California.

During components of the students' math, science, social studies, and health classes, the garden is abuzz as the kids literally get their hands dirty learning by doing. The fruits of their labor then become snacks for them and their classmates. Not only are the kids learning about foods they're used to seeing on their plates, their education includes learning about native Hawaiian species and canoe plants—those plants early Polynesian settlers felt were so life sustaining, they carried them clear across the ocean to take root in their new island home.

Volunteers can be a part of this revolution. On Saturday mornings, every four to six weeks, Māla'ai hosts community workdays, where volunteers, students, and community members gather for a garden work party (which may include building or maintenance projects). For volunteers able to make a longer-term commitment, there's help needed working with kids in the garden during school hours each week (or on your own time), and with the ongoing tasks of administration and fund-raising.

"When my son started at the school, I thought it'd be a great way to be involved with the school and the kids. I really believe in hands-on education. A lot of kids don't do well in the classroom, and the garden's a great place to learn.

"I love the kids. It's fun to see learning in action, and the excitement they get from learning something new, the sense of pride they feel.

"It's amazing to see in the modern age how little kids know about their food, about where vegetables come from. It has definitely heightened my son's interest in plants and planting. He's much more aware now of plants in the neighborhood. We found a native Hawaiian cotton plant in a disturbed field, where they had used Roundup, and he said, 'Mom, we have to rescue that plant.' So we dug it up and brought it home.

"It really connects the kids to the Hawaiian culture too. They grow native plants, the Hawaiian class comes out and chants, they talk about planting with the moon. . . .

"Having my own little garden, it's been great to see what a lot of hands can do."
—Holly Sargeant-Green
Waikoloa, Hawai'i Island

Volunteer Activities: Community garden workdays, weekly gardening with middle-school students (or on your own), administration/fund-raising.

When: Community workdays—Saturday mornings, every 4 to 6 weeks, for 3 hours (see Web site for upcoming dates). Weekly gardening—during school year, 10-week commitment requested, 1 day per week; summer garden maintenance also needed. Administration—ongoing.

Who: Community workdays—both individuals and small groups welcome. (Young children must be supervised by an adult.) Weekly gardening—individuals. Administration—individuals with basic office and/or fund-raising skills.

Hardiness Level (5 being most difficult): 1 to 4

Advance Notice: Community workdays—none needed for individuals; contact regarding small groups. Weekly gardening or administration—contact to schedule an appointment.

Donations: Monetary donations are appreciated and tax deductible. Contact regarding equipment and supplies needs.

Contact: E-mail preferred.
Amanda Rieux
Māla'ai
The Culinary Garden of Waimea Middle School
67-1229 Māmalahoa Highway
Kamuela, HI 96743
(808) 887-6090
iamrieux@speakeasy.net
www.malaai.org

Mokupāpapa Discovery Center

Imagine Hawai'i before the islands were inhabited by people: Warm sand beaches teeming with tropical seabirds. Monk seals stretched out in the sun. The oceans flush with colorful fish and graceful green turtles. Coral reefs as vivid and diverse as any aquarium could imitate.

Virginia Branco—NOAA/PMNM

This is the nature of the Northwestern Hawaiian Islands (NWHI), an oasis several hundreds of miles northwest of Kaua'i. Uninhabited by humans for more than two hundred years—many islands never having seen human settlements—the region has earned the nickname of America's Galápagos. Native Hawaiians call them sacred, considering them *kupuna*, or ancestor, islands.

When international trade and commerce came to Hawai'i's waters in the eighteenth and nineteenth centuries, the islands were sadly exploited. Seabirds were harvested for their feathers, sometimes their eggs. Islands were mined for guano, to be processed into fertilizer. Entire ecosystems disappeared. And although the area is protected now as the Papahānaumokuākea Marine National Monument, illegal activities such as dumping and poaching still occur.

Legal access is limited to research, Hawaiian cultural practices, and other permitted activities, so those of us who want to experience the islands' treasures head instead for Mokupāpapa, the NWHI discovery center at Hilo Bay. Interactive displays, a 2,500-gallon aquarium, and multimedia exhibits paint a three-dimensional picture of the islands, and bring their swimming and soaring creatures to vibrant life.

Volunteers can get involved with all kinds of projects working alongside an experienced volunteer—answering questions and directing the public, leading school groups, developing and implementing lesson plans, maintaining the aquariums, or helping run the volunteer program itself. There's also a need for monthly guest speakers by university researchers, marine conservationists, and cultural practitioners.

> "My love of the ocean is a natural fit for volunteering at Mokupāpapa, the discovery center for our new marine national monument. I have been here since the center opened in 2003 and enjoy meeting visitors, explaining the formation of our archipelago, and describing the wildlife that lives on its remote atolls."
> —Patricia Richardson
> Hilo, Hawai'i Island

Volunteer Activities: Public education, program development and leadership, aquarium maintenance, guest lectures.

When: Year round, 1-month minimum commitment.

Who: Individuals and small groups (up to 3 people) welcome, age 18 and older. A background in marine education or conservation is highly desirable, but not required.

Hardiness Level (5 being most difficult): Public education, program development and leadership, guest lectures—1. Aquarium maintenance—3.

Advance Notice: Public education, program development and leadership, aquarium maintenance—2 weeks. Guest lectures—2 months.

Education: As an educational center, public programs are free and ongoing. Additional free summer programs are available for school groups. (Contact for details.)

Donations: Monetary donations are appreciated and tax deductible if made to the National Marine Sanctuary Foundation (go to NMSF Web site at www.NMSFocean.org).

Contact: E-mail preferred.
Yumi Yasutake
Mokupāpapa Discovery Center
Papahānaumokuākea Marine National Monument
308 Kamehameha Avenue, Suite 203
Hilo, HI 96720
(808) 933-8195
yumi.yasutake@noaa.gov
www.hawaiireef.noaa.gov

Na Ala Hele—Hawai'i Island

Imagine lava roiling from the mouth of a volcano on its course toward the waters of the waiting sea. It encounters a forest, a pristine gem of nature, where endangered birds flit in and out of the upper canopy, a thriving panoply of native plants beneath. . . . The lava decides to spare the forest. It curves in an arc around either side of the lush vegetation, leaving an untouched oasis—a *kīpuka*—in its wake.

Kaulana Manu ("safe haven for birds"), formerly known as Kīpuka 21, is one of these marvels of nature. Along the Saddle Road, the lava-ringed ecosystem boasts *koa* and *'ōhi'a lehua* trees rising up to sixty feet, while *'ākepa*, *'i'iwi*, and *'ōma'o* wings flutter in the wind. It's raw and rugged—that's what makes it so special. And that's what makes for lots of work to come before its restoration is completed.

Volunteers with Na Ala Hele on Hawai'i Island can help by removing alien plants and weeds, and by maintaining sections of the trail.

Greg Lindstedt

"Kīpuka 21 has long been a favorite location of bird seekers to find native Hawaiian forest birds, probably ever since reports from bird monitoring projects noted that many native forest species were using these small oases in old lava flows. . . .

"The work has been very fulfilling because I got to do things that I had never had an opportunity to do before. I had helped to cut trails in Castle Rock State Park in the Santa Cruz Mountains of the San Francisco Bay Area, but building and improving trails in a forest grown on lava was a very unique experience. Additionally, the location abounds with great examples of native Hawaiian vegetation.

"A satisfied smile spreads across my face every time I take a break to wipe the sweat from my brow and I hear an *'ōma'o* break into song, or see the vermilion flash of an *'i'iwi* dashing through the canopy. It's the best kind of work anywhere."

—Les Chibana
Volcano, Hawai'i Island

There's also a need for resident volunteers skilled in ecology, botany, or bird identification to lead educational hikes through the area.

(See Multi-Island section, under Na Ala Hele, for more about the agency itself.)

Volunteer Activities: Trail maintenance, removal of invasive species, leading educational hikes.

When: Year round. (Contact for upcoming dates.) Free native versus invasive plant tours given to all volunteers.

Who: Both individuals and groups welcome. Kids under 8 must be supervised by an adult. Potential hike leaders must be skilled in their subject and attend several hikes first, then obtain approval for their proposed tour.

Hardiness Level (5 being most difficult): 1 to 5

Advance Notice: Trail work—1 day. Leading educational hikes—1 week.

Education: Free public tours are given twice monthly on native versus invasive plants and birding. (Contact for upcoming dates and fees.)

Donations: Monetary donations are appreciated and tax deductible. Contact regarding equipment and supplies needs.

Contact: E-mail or phone.
Virginia Aragon
Division of Forestry and Wildlife
Na Ala Hele—Hawaiʻi Trail and Access System
19 East Kāwili Street
Hilo, HI 96720
(808) 974-4221
varagon@dofawha.org
www.hawaiitrails.org

Sierra Club—Moku Loa (Hawai'i Island)

Volunteering with the Sierra Club on Hawai'i Island might take you to Kalōpā, a mountainside forest with an upper canopy of native 'ōhi'a trees, or to a historic seaside trail in Puna, or to Hakalau on a conservation trip with the Fish and Wildlife Service. You might find yourself clearing trails, weeding, planting native species, or simply cleaning the place up.

(See Multi-Island section, under Sierra Club—Hawai'i Chapter, for more about the organization itself.)

Volunteer Activities: Removing invasive species, replanting natives, trail clearing and maintenance, cleanups.

When: Year round, project lengths vary.

Who: Free for Sierra Club members, nominal fee for nonmembers, unless otherwise noted on Web site. Age 12 and older; kids under 18 must be supervised by an adult.

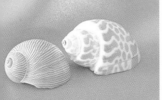

Hardiness Level (5 being most difficult): 1 to 5

Advance Notice: See Web site.

Education: Outings such as hikes along Chain of Craters Road in Hawai'i Volcanoes National Park, along the arid South Kohala Coast to view petroglyphs and rare anchialine ponds, or to isolated backcountry beaches. (See Web site for details and fees.)

Donations: Monetary donations are appreciated and tax deductible if made to the Sierra Club Foundation (contact office for more information).

Contact: See Web site for particular project coordinator's contact.
Sierra Club—Moku Loa Group
P.O. Box 1137
Hilo, HI 96721-1137
(808) 965-5460
campbellcenter@hawaiiantel.net
www.hi.sierraclub.org (click on Moku Loa Group)

Sarah Moon

Three Ring Ranch Exotic Animal Sanctuary

Dogs and cats have been bred to be good companions. We domesticated them thousands of years ago, selecting traits through the generations that made sharing our lives with these animals easier.

Living with a zebra isn't quite so easy. Nor is living with a cheetah, once it grows past that cute, cuddly kitten stage.

The exotic animal trade is rampant, even here in Hawai'i, and much of it illegal. Then when the pets become too large, too strong, too unmanageable, owners often desperately want to give them away. Other times, exotic pets escape into our fragile ecosystem and are picked up by local authorities.

Many of these creatures stay at Three Ring Ranch until they can be relocated, or become permanent residents. The sanctuary welcomes all and asks no questions, so they also serve as an amnesty center—meaning anyone can anonymously drop off an alien or injured animal. The ranch will do the rest.

Home to zebras, flamingos, and peacocks, Three Ring Ranch also houses and rehabilitates native wildlife, such as the *'ōpe'ape'a* (Hawaiian hoary bat), *pueo* (Hawaiian short-eared owl), *nēnē* (Hawaiian goose), and many species of birds. Short-term volunteer opportunities are for groups only, where you may spend the day working in the wetlands habitat for *nēnē*, or keeping the pastures clear of wild growth, or helping with construction on the barn.

For volunteers who commit two months or more, there are internships available, where you can become a daily animal keeper, learning from sanctuary curator Ann Goody not just how to caretake a wild creature, but how to communicate with it in its own language. As a wildlife "whisperer" herself, Ann's personal teaching philosophy provides a rare volunteer opportunity.

Volunteer Activities: Wetlands upkeep, pasture work, construction on barn; animal keeper internships; data entry (contact for details).

When: Upkeep and maintenance—year round, for 1 day per project. Internships—2-month minimum commitment, for 3 hours per day.

Who: Groups welcome for day projects; individuals for internships. Kids must be age 17 or older (unless part of a school group). (Resident volunteers who live in or near Kona are always welcome.)

Hardiness Level (5 being most difficult): 3 to 4

Advance Notice: Groups—1 month (contact by e-mail); individuals—apply up to February 15 for the following summer (see Web site for details and online application).

Education: The sanctuary is not open to the public, but a wide range of educational programs are available. (See Web site or contact for details.)

Donations: Monetary donations are appreciated and tax deductible. Contact regarding equipment and supplies needs. Ask about the Adopt-an-Animal program.

Contact: E-mail or Web site.
Ann Goody
Three Ring Ranch Exotic Animal Sanctuary
75-809 Keaolani Drive
Kailua-Kona, HI 96740-8815
(808) 331-8778
animals@threeringranch.org
www.threeringranch.org

"I feel blessed to have had the chance to work with Drs. Ann and Norm Goody at their Exotic Animal Sanctuary. There I have received exposure to animals and insight on how a wildlife facility is run. . . . I learned, most importantly, that it is paramount to listen to the animals and if you pay close enough attention they will always tell you exactly what they need.

"Volunteering at the Three Ring Ranch has helped me to realize that my passions can also be my career. . . . By the end of the internship I was fully aware of the commitment, responsibility, and focus required to care for these amazing and wild creatures."

—Kirsten Stambaugh
Humboldt, California

"Living at the ranch was extremely valuable as Dr. Ann Goody helped us with a 'hands-on' style of instruction with the animals. Oftentimes before dinner, we would listen to Dr. Goody's personal stories, examine museum specimens, or participate in wildlife calls. . . .

Three Ring Ranch

"There are few places where one can interact directly with zebras, oryx, and all manner of birds and small exotic animals in one day, not to mention over a span of three weeks. Fewer still are instructors both willing and able to impart their knowledge of how to communicate with animals. . . .

"Three Ring Ranch has changed the way that I interact with animals of any size, shape, or temperament."

—Allison M. Cleymaet
Berkeley, California

TREE (Tropical Reforestation and Ecosystems Education) Center Hawai'i

TREE is rebuilding forests, one planting at a time.

Sometimes it's the *loulu* fan palm, the only palm native to Hawai'i—important to dryland forests along the coast. Or it might be the native *koa* tree, a cultural symbol throughout the islands, being replanted in a montane (moist, cool, upland) forest on Mauna Loa's slopes. In a mesic (moderately moist) forest above Kona, TREE participants have planted over three thousand *koa* seedlings.

Education with TREE is always a hands-on experience. Their youth programs take kids out on field days, nature camps, and exchange programs with native cultures from other states, such as the Yupik and Inupiat Eskimo or Navajo Indian. Adult education includes classes at the Kona Community School for Adults, and getting your fingers in the dirt in the plant nursery and at restoration sites.

Volunteers can take part in this learning while doing. By removing invasive species, propagating natives, and replanting them back in the forest, you'll discover firsthand the intricate balance of the forest ecosystem. Often

TREE Center Hawai'i

just clearing away nonnative plants, eliminating the competition for nutrients, sunlight, and rain, allows native seedlings to naturally reappear.

The native forest is a rich, concentrated resource. Helping to restore it can lay the foundation for an entire island ecosystem.

Volunteer Activities: Removing invasive species, propagating and planting natives, clearing trails.

When: 6 times per year, for 2 to 3 hours.

Who: Both individuals and groups (up to 12 people) welcome, age 9 and older.

Hardiness Level (5 being most difficult): 2 to 4

Advance Notice: 1 month.

Education: Extensive environmental education programs for youth include field days, spring and summer camps, cultural exchange camps, and classroom nursery programs. Adult education includes horticulture and habitat restoration training. (See Web site or contact for details and fees.)

Donations: Monetary donations are appreciated and tax deductible. Contact regarding equipment and supplies needs.

Contact: E-mail or phone.
Christy Schumann
TREE (Tropical Reforestation and Ecosystems Education) Center Hawai'i
P.O. Box 411
Kealakekua, HI 96750
(808) 333-0330
info@treehawaii.org
www.treehawaii.org

"When [my daughter] came home [from camp], after we talked about all the things she did and enjoyed, she rolled out her Pu'u Wa'awa'a poster, studied it, and asked me if she could put the poster on the wall next to her bed!

"To me, [her] wanting that poster up about the endangered plants, animals, and forest, [and] comparing invasive species of plants, vegetation, and animals, shows me that TREE Camp 'got the message' across to my daughter. She really understands now that our island(s) are very fragile."

—Robin M. Aweau
Hāwī, Hawai'i Island

"I feel it's important as adults that we need to learn and care for our land now, so we can teach our children the importance of stewardship for the sake of their future. . . .

"I couldn't volunteer [when I lived in] Los Angeles. People would ask, Why are you helping? Why are you helping that person move? It took me coming all the way to Hawai'i to be able to give and not be expected to get back."

—Tracey Seki Matsuyama
Hōnaunau, Hawai'i Island

The United Nations headquarters in New York City has also been celebrating Earth Day for nearly forty years with their own ceremony, but they've chosen the spring equinox (around March 20) to commemorate the day. At the precise moment of the equinox, they ring the Japanese Peace Bell (donated by Japan to the United Nations), a reminder of the importance of preserving and caring for the earth's natural resources and all forms of life.

The April 22 date is often associated with the lifetime work of well-known conservationist John Muir (1838–1914), due to its proximity to his birthday, April 21. Responsible for establishing our national park system and many other environmental achievements, the Wilderness Prophet, however, has his own holiday: John Muir Day was federally designated in 1988 as April 21.

Every Day Is Earth Day

For the individuals featured in this book, every day truly is Earth Day. A day to listen to the islands and their wild creatures, to understand their plights through the hearts and minds of the people determined to protect them.

On April 22, 1970, the first national Earth Day brought out more than twenty million people across America. Initiated by a U.S. senator from Wisconsin, Gaylord Nelson, this demonstration of concern over the environment soon spurred new legislation—the Environmental Protection Agency was established that year; the Clean Air, Clean Water, and Endangered Species acts within the few years that followed.

Earth Day has since grown into a global environmental celebration, rallying more than five hundred million people annually in 175 countries. In Hawai'i, it has many faces—university and college campuses hold community festivals, musicians and performers share the stage with ecologists and activists. There are lectures, movie screenings, even educational boat tours. And everyone shares a collective message: the necessity of environmental stewardship.

Many organizations in this book have their own Earth Day extravaganzas—you might spend the day planting native trees with TREE Center Hawai'i, or clearing marine debris from the coastline with Hawai'i Wildlife Fund, or helping out at a booth at Maui Nui Botanical Gardens to educate others about the islands' endangered species.

Some organizations also celebrate national events like Arbor Day—groups such as the Amy Greenwell Ethnobotanical Garden give away native trees each November to the first lucky folks to claim them. During the International Coastal Cleanup in September, Project S.E.A.-Link and Reef Check Hawai'i take extra care of the islands' fragile shores.

Still other groups host their own brand of annual festival, like the Hawaii Audubon Society's Christmas Bird Count, the Lyon Arboretum's 'Awa Festival (in honor of the traditional ceremonial plant), or Community Work Day Program's "Get the Drift and Bag It" program. Contact the organizations you're most drawn to and find out what's happening right now.

"Everybody needs beauty as well as bread, places to play in and pray in, where Nature may heal and cheer and give strength to body and soul alike."

—John Muir, *The Yosemite* (1912)

Peter S. Goltra for the National Tropical Botanical Garden

Found only on the island of Kaua'i, this native hibiscus, Hibiscus waimeae, is one of the Pacific's many jewels.

KAUA'I

Photo by Mary Shuford for the
National Tropical Botanical Garden

Friends of Kamalani and Lydgate Park

You don't have to be a kid to be awed by the Volcano Slide at Lydgate Park's Kamalani Playground. Or by the multicolored mosaics—turtle tables and snake seats and tiles shaped into the faces of divers and butterflies and fish. Or by the whimsical sculptures that create a fantasy forest on the Kamalani Kai Bridge.

U.S. Navy photo by Johnny Michael

A lot of love went into creating what the Friends of Kamalani and Lydgate Park like to call the "best beach park in the state of Hawai'i." And love is what keeps it looking spectacular. Major workdays in the spring and fall turn out at least one hundred volunteers to help maintain the beach park's community-built structures. And once a month, volunteers help clear litter from the beach and remove the debris that washes down from the Wailua River after a heavy rain.

The many people responsible for this beautification effort have formed lasting relationships—among themselves and with the land itself. In the words of volunteer Tammi Andersland, "It was volunteering at Kamalani and Lydgate Park that helped our family get connected to the community. . . . We looked forward to our Saturday mornings and the building of the [Kamalani Pavilion]. Our kids met other children on the build site and really started to feel connected to the island."

Volunteer Activities: Beach cleanups and beautification, maintenance of beach park structures.

When: Beach cleanups—first Saturday morning of every month, for 2 hours. Maintenance of structures—Saturday nearest Earth Day (April 22) in the spring, and National Make a Difference Day in the fall (fourth Saturday in October).

Who: Both individuals and groups welcome. Youth must be age 18 or older to use power tools, with demonstrated proficiency. Kids under 12 must be supervised by an adult.

Hardiness Level (5 being most difficult): 1 to 3

Advance Notice: Needed for groups only.

Donations: Monetary donations are appreciated and tax deductible if made through YWCA (see contact information below).

Contact: E-mail preferred.
Thomas Noyes
Friends of Kamalani and Lydgate Park
c/o YWCA of Kaua'i
3094 'Ēlua Street
Lihu'e, HI 96766
(808) 245-5959
thomasnoyes@hawaiiantel.net
www.kamalani.org

"Working on Kamalani Playground, in various beach cleanups, building Kamalani Kai Bridge and painting Kamalani Playground wood with linseed oil, and helping at Habitat for Humanity build-a-thons were all wonderful experiences.

"I am not very strong physically (I'm a sixty-four-year-old woman), but in all cases there was always something fun and easy to do, from staffing the sign-in table to minding children in day care to serving lunch to bringing cold water to the Habitat folks who were doing the heavy lifting. . . .

"What I have learned is that the nicest people in the world pitch in and help with these sorts of events. Without exception the 'work' days have been filled with smiles, laughter, joy, cooperation, good times, and a very strong sense of positive accomplishment. The 'work' has actually been more fun in some cases than what normally gets classified as 'recreation.' . . .

"At the end of each 'work' day, I have been tired, but gloriously happy."

—Mary Mulhall
Kapaʻa, Kauaʻi

"I am a regular swimmer at Lydgate Pond and a regular beach walker/runner at Kitchens, and see both as treasured recreational/fitness resources for the east side of Kauaʻi. I want to help preserve and maintain these unique coastal areas for us all, so I pitch in on construction (Kamalani Bridge), maintenance (Kamalani Playground), and cleanups (Hikinaakalā Heiau and the beach) when I can.

"There is always good cheer, good teamwork, and the refreshing dip in the ocean, and often a shared meal *pau hana* [after work]. Early mornings on the east side are glorious with the cool trade winds and low sun."

—Linda Pizzitola
Wailua Houselots, Kauaʻi

Hanalei Watershed Hui

Water is life. It's the "blood of the watershed," in the words of Kīlauea Elementary School student Ross Kimura. It links the summit to the sea, and sustains all the creatures in between—a good reason to make sure the water's healthy.

This is the work of the Hanalei Watershed Hui (HWH). Devoted to ensuring the quality and flow of the streams and basins leading to the Hanalei River and Bay, they recognize the value of protecting this precious resource. That includes reducing erosion from the riverbanks and trails that drain into Hanalei's waterways. Volunteer projects might include planting ground cover, filling in gullies, and installing erosion-control matting. Or you might be testing water quality, doing agricultural or fishpond work, or helping staff a mobile display at special events, encouraging others to take stewardship of local water sources.

Photo courtesy of Hanalei Watershed Hui

Through the Hui's educational programs, they also reach out to Kaua'i's children, helping them make connections between sky and land, land and sea. And the kids are getting it. As Kula Elementary School student Natalia Smith describes, "Now I see that if we don't help to prevent erosion, we will have to surf in murky water when it rains."

Nobody wants to surf in murky water, nor can the creatures who live in the water survive in a murky home. Becoming more intimate with our water will help ensure its ongoing health.

Volunteer Activities: Water quality monitoring, trail maintenance, special projects (such as fishpond restoration, agricultural work, or staffing a mobile display).

When: Year round, for 1 to 3 hours per project.

Who: Individuals preferred; small groups welcome. For water quality monitoring—age 12 and older. For trail maintenance—age 14 and older.

Hardiness Level (5 being most difficult): 3

Advance Notice: 1 month.

Education: They have a lending library of videos on all things relating to water and the Hawaiian *ahupua'a* (ancient land division).

Donations: Monetary donations are appreciated and tax deductible. Contact regarding supplies needs.

Contact: E-mail preferred.
Makaala Kaaumoana
Hanalei Watershed Hui
P.O. Box 1285
Hanalei, HI 96714
(808) 826-1985
hanaleiriver@hawaiian.net
www.hanaleiwatershedhui.org

"Rain or shine the work on 'Ōkolehao Trail lifts my heart! The vistas are breathtaking and the trail is a good workout without being too challenging to discourage repeat visits. I have met people who use the trail daily or weekly as part of their health regime.

"It is very satisfying to be aligned with HWH . . . truly a worthy organization! What a great way to enrich my visit to Kaua'i!"

—Jesi Trego
Santa Fe, New Mexico

"My volunteer work has greatly enriched my enjoyment of the 'Ōkolehao Trail. It has been an unexpected pleasure to work alongside scientists in the field as well as dedicated members of the community.

"I have also observed the enthusiastic participation of local schoolchildren who have had the opportunity to plant native plants alongside the trail. They will be able to take pride in their efforts as they return to the trail throughout their lifetime."

—Lauryn Galindo
Princeville, Kaua'i

Wendy McIlroy

Hawaiian Monk Seal Conservation Hui

When you volunteer with the Hawaiian Monk Seal Conservation Hui, you become a voice for these sleek creatures. You become an educator of their behaviors and habitat. You become the go-to person for passing beachgoers wanting to learn about the Hawaiian monk seals' precarious existence, and the need for humans to help ensure their survival.

In return, you get to experience a seal's daily life. You might be privy to the interaction between mother and pup, or witness to a newborn's first few hours. Sometimes you'll act as seal sitter, protecting the seal from curious onlookers, as it must rest on the sun-warmed sand. To simply be able to observe these remarkable creatures is reward in itself.

The Hui also regularly hosts beach cleanups to keep the seals' onshore habitat pristine. And they respond to the occasional emergency, such as untangling a seal from a fishing line and hook or net.

As the Hawaiian monk seal is one of the most endangered of all seals, these marine mammals need the support of those who share their ocean home.

Volunteer Activities: Monk seal monitoring and public outreach, beach cleanups.

When: Year round, for 2- to 3-hour shifts. (Free volunteer training for monk seal monitoring is done on-site or at bimonthly evening meetings.)

Who: Both individuals and groups welcome, age 8 and older. (Kids under 15 must be supervised by an adult.) For seal monitoring, group limit is 3 people; for beach cleanups, larger groups welcome.

Hardiness Level (5 being most difficult): 1 to 3

Advance Notice: Seal monitoring—none needed (see above regarding training). Beach cleanups—at least 2 weeks.

Education: Public education occurs at monthly events around the island.

Donations: Monetary donations are appreciated and tax deductible if made through the DLNR, NOAA Sanctuary, or Hawai'i Wildlife Fund. (Signify the donation is for Hawaiian monk seal conservation—contact the Hui for more details.) Contact regarding equipment and supplies needs.

Contact: E-mail or phone.
Mimi Olry
Hawaiian Monk Seal Conservation Hui
Division of Aquatic Resources/DLNR
3060 'Eiwa Street, Room 306
Līhu'e, HI 96766
(808) 346-1592
hawaiianmonkseal@msn.com

Wendy McIlroy

Hawai'i DLNR/DAR

"It isn't about being endangered to me, it's my sense of being responsible for the interference we've made on other species—and I'm part of it. This is a small thing I can do to assuage that sense of responsibility.

"I knew volunteering involved commitment, passion, dedication. I knew it was based on intent. That is all I care about: observing an animal, documenting its assistance, and then educating anybody that I might attract by my presence there.

"Whatever it takes, and whenever—it's a priority for me. I have other lives, but if I get the call, everything else takes second."

—Donna Lee
Waimea, Kaua'i

"I did a lot of community work when I was in business, but it didn't have to do with the environment. It had to do with people. . . . I still have my human resources background, and I talk to people a lot. Most rewarding to me is that people go away with a much better understanding of what the seals are facing—even though their overall population is dwindling, more people are aware of it.

"Those creatures that can't speak for themselves, you can be a voice for them."

—Larry Rauchut
Kalāheo, Kaua'i

Hui o Laka
Kōke'e Natural History Museum

Leave the highway and head up the mountain toward Kōke'e, and you leave time, entering a realm like none other in the islands. In vivid contrast to the tropical playground below, Kōke'e resides in cool calm at around four thousand feet, floating among the clouds. Heavily forested, it offers spectacular bird's-eye views from strategic overlooks into the canyons and valleys below. On the mountain side, it flanks the Alaka'i Swamp, which stretches farther inland to meet up with Mount Wai'ale'ale, one of the wettest spots on earth.

Kay Koike

For over fifty years, Hui o Laka has run the Kōke'e Natural History Museum—the on-site resource for visitors wanting to understand more about this rare island environment. The staff greets over one hundred thousand visitors each year. In the in-between hours, you can find them "forest gardening," helping a much-underfunded state parks system bring the area's trails, roadsides, and overlooks to the weed-free condition such a precious ecosystem deserves.

As a volunteer with Kōkua Kōke'e (what Hui o Laka calls its forest stewardship program), you might end up collecting seeds or trimming branches, weeding trails or restoring historic structures. In exchange, the spirit of Kōke'e is sure to become a part of you.

Volunteer Activities: Forest gardening and preservation of historic structures.

"I've been in the Peace Corps and in a lot of different countries. That takes a lot of learning about and integrating yourself within the culture. Here on Kaua'i there are different cultural groups, and each group more or less stays with their own culture. But up in Kōke'e, the different groups and cultures mix.

"I help with the setup and teardown of events. And I maintain the sidewalk that goes around the square. I do those things I felt nobody else was doing. I like to work independently—it's almost like a meditation for me. And at different times it's allowed me to stay in the cabin.

"Being in nature is getting hard to do now, with all the development. I really like the mountains and I really like the beach. And the nights up at Kōke'e, when there's no one out there, and it's pitch-black, and the stars are out—that's reward enough for me."

—Gary Hoover
Kekaha, Kaua'i

When: Year round, for 1 day per project.

Who: Both individuals and groups welcome, age 5 and older. (Children must have a parental consent form.)

Hardiness Level (5 being most difficult): 1 to 3

Advance Notice: 1 to 2 days.

Education: A 4- to 7-week Saturday field lecture series occurs each spring. (Contact for details and fees.)

Donations: Monetary donations and memberships are appreciated and tax deductible. Contact regarding equipment and supplies needs. Ask about sponsoring a *kumu hula* (*hula* teacher) to travel to their annual Emalani Festival.

Contact: E-mail, phone, or postal mail.
Michelle Hookano
Hui o Laka
P.O. Box 100
Kekaha, HI 96752
(808) 335-9975, extension 22
kokeemuseum@earthlink.net
www.kokee.org

Kay Koike

Kōke'e Resource Conservation Program

The state parks of northwest Kaua'i are vast—over twelve thousand acres of uninhabited wilderness. Like other places in the islands, they're being threatened by invasive species, and there's lots to be done to maintain the region's natural state.

Set up in 1998 as a program of Hui o Laka (see previous listing), the Kōke'e Resource Conservation Program specifically tackles the overgrowth of unwanted species. As a volunteer, you'll get to spend some time in the area—eradicating invasives by day, bunked up in the volunteer cabin by night.

A few days' work then affords you a few days off during your stay. You'll have time to explore the trails that run like rivulets through nearby Waimea Canyon, which Mark Twain famously called the Grand Canyon of the Pacific. Or you might head for Alaka'i Swamp, a boggy, otherworldly trek toward Kaua'i's highest point, Mount Wai'ale'ale.

Maybe you'll find yourself at the end of the climbing mountain highway, where, on a cloudless day, you can peer over the cliffs and into Kalalau Valley, surely one of the most stunning sights in all the Pacific.

Kōke'e Resource Conservation Program

If the timing's right, you might have the cabin to yourself. And since park staff live off-site, volunteers must be self-motivated with a love for being on their own.

Volunteer Activities: Treatment of invasive species. *(This typically involves applying over-the-counter herbicides—if you prefer to not use herbicide, the staff will design a volunteer project for you involving only pulling or cutting.)*

When: Year round, flexible time commitment. (Lodging available in volunteer cabin.)

Who: Both individuals and groups welcome. Kids under 17 must be supervised by a parent or approved chaperone.

Hardiness Level (5 being most difficult): 1 to 5

Advance Notice: As much as possible.

Donations: Monetary donations are appreciated and tax deductible if made through Garden Island Resource Conservation and Development (contact office for more information). Contact regarding equipment and supplies needs.

Contact: E-mail or phone.
Katie Cassel or Tammy Goodall
Kōke'e Resource Conservation Program
P.O. Box 1108
Waimea, HI 96796
(808) 335-0045
rcp@aloha.net
www.krcp.org

Kōke'e Resource Conservation Program

"Kōke'e was like an extended backyard while I was growing up, since I lived/live in Kekaha, right down the mountain. Even when coming home from school vacations, I always had to go up to Kōke'e at least once. . . .

"It's beautiful whether it's sunny, foggy, or rainy. Many of us feel the *mana*/spirit of the Kōke'e mountains and what it can do for one's soul or being. . . .

"From the 'extended backyard'/playground for family and school events, it has become a place of great value that can't be expressed in dollar amounts; like in the credit card TV commercial, it's priceless.

"[The most rewarding part has been] satisfaction in knowing that I can help to keep Kōke'e native where it's possible, one weed at a time; that the mostly green of the 'shy' and less colorful native flora is more valuable and beautiful than the showy plants of the invasive species—like the princess or glory flower, banana poka, yellow and *kāhili* gingers, etc."

—Kay Koike
Kekaha, Kaua'i

"From my arrival and during the duration of my stay, I've felt like there has always been an extraordinary effort to make myself and the other volunteers feel welcome. I always felt like I was treated as part of the *'ohana* [family]. . . .

"The program content was very interesting and I learned a ton . . . everyone provided a wealth of information about the local flora and fauna, and were accommodating of my continued questions."

—Mark Wasser
Honolulu, O'ahu

Sierra Club—Kaua'i

Kaua'i has the most trails of any Hawaiian Island—from mountainous hiking paths at Kōke'e and Waimea Canyon state parks to the shores of Māhā'ulepū, a wild and fragile coastline on the island's south shore. "Each trail has its own special features and all have one thing in common—gorgeous scenery!" says Sierra Club Kaua'i volunteer outings coordinator Judy Dalton.

To help protect this cherished environment, the Kaua'i Group hosts at least one cleanup per month. Sometimes they even take to inland waters, where they forge the American Heritage Hanalei River by kayak, for example, cleaning up debris and beautifying the river's banks.

(See Multi-Island section, under Sierra Club—Hawai'i Chapter, for more about the organization itself.)

Volunteer Activities: Cleanups at beaches, rivers, and roadsides.

When: At least 1 cleanup per month, for 2 hours.

Who: Both individuals and groups welcome, all ages.

Hardiness Level (5 being most difficult): 1

Advance Notice: See Web site.

Judy Dalton

Education: Outings such as sunset whale-watching adventures, full-moon hikes through the National Tropical Botanical Garden, or treks along a forested river laced with waterfalls. (See Web site for details.)

Donations: Monetary donations are appreciated and tax deductible if made to the Sierra Club Foundation (contact office for more information).

Contact: E-mail or phone.
Judy Dalton
Sierra Club—Kaua'i Group
P.O. Box 3412
Lihu'e, HI 96766
(808) 246-9067
dalton@aloha.net
www.hi.sierraclub.org (click on Kaua'i Group)

"I got involved because I felt I could help the environment. I could make people aware of endangered and endemic Hawaiian plants by getting them outside to see what the true environment is like, where there are not a lot of alien plants. When we go out to Kōke'e or the Nāpali Coast, where there are still native plants, people are amazed.

"Every hike is a new experience. You might see something differently than you've ever seen it before."
—Jane Schmitt
Princeville, Kaua'i

Surfrider Foundation—Kaua'i

Bottle caps, plastic bags, lighters, batteries. This isn't the natural diet of a seafaring bird or marine mammal, but it's increasingly becoming their standard fare. Many creatures tragically ingest these harmful items, mistaking them for food.

Marine debris is an enormous issue for wildlife whose lives

Surfrider Kaua'i

depend on the sea, so Surfrider Foundation on Kaua'i devotes much of its time to cleaning up the island's beaches. There's even a dedicated net patrol to tackle the massive entanglements that often wash up on shore—nets can break off coral or block sunlight to the underwater ecosystem; sometimes they wrap around sea turtles or monk seals, even whales.

Surfrider volunteers also make up the Blue Water Task Force, paddling out to favorite surf spots to test for water clarity, salinity, and bacteria. Then there are special events, like when Surfrider and Barefoot Wine teamed up to comb the sands of East Kaua'i to ensure they'd be barefoot friendly. Volunteers are the sole force behind these valuable efforts.

(See Multi-Island section, under Surfrider Foundation, for more about the organization itself.)

Volunteer Activities: Beach cleanups, Net Patrol, Blue Water Task Force, special events.

When: Beach cleanups, Net Patrol, Blue Water Task Force—monthly, for 1 to 3 hours. Special events—quarterly.

Who: Both individuals and groups welcome, all ages.

Hardiness Level (5 being most difficult): Beach cleanups—1. Net Patrol—3. Blue Water Task Force—4 (must have surfboard and be comfortable paddling).

"It's rewarding working with the other groups on Kaua'i. We're not just alone. We're in collaboration with many other like-minded groups.

"The first time we went out to pull a net, there were four Surfrider volunteers and three more helpers off the beach. The net was probably five times each of our body weights. It was huge! In one hour we did it—so just a few people can make a difference.

"It's a great way to be in service and do love in action, which is what I think volunteering is—something that you love and you're in action doing it. It's keeping Kaua'i beautiful, and being part of the solution instead of the problem."

—Barbara Wiedner
Kapa'a, Kaua'i

"I was raised in a family where we were taught to serve and give back to our community, so it's natural for me to want to pass that value on to my own children. We wanted both of them to really understand how fragile our ecosystems are and to extend what they are learning in school. . . .

"I am amazed at what we find on the beaches. . . . I could not believe what we found at the end of Keālia . . . batteries, refrigerator parts, shoes, furniture, etc. It was disheartening. . . .

"We pick up everything we can see. We get to chat and discuss what we are finding, and we play games where we imagine what kind of person/family or group would have left their trash, or how items could have washed up on the beach. We also talk about the different items and how they could damage the reef and marine life. It's fun—there is a great sense of camaraderie and a wonderful feeling of accomplishment when you are done."

—Lisa Mireles
Kīlauea, Kaua'i

Advance Notice: Beach cleanups—none needed. Net Patrol—1 week. Blue Water Task Force—1 month.

Education: Keynote speakers talk on marine conservation topics at free quarterly meetings.

Donations: Monetary donations are appreciated and tax deductible. Contact regarding equipment and supplies needs.

Contact: E-mail preferred.
Sheri Saari
Surfrider Foundation, Kaua'i Chapter
P.O. Box 936
Kilauea, HI 96754
(808) 652-4648
surfriderkauai@gmail.com
www.surfrider.org/kauai

Waipā Foundation

When the North Shore *ahupua'a*, or traditional land division, of Waipā was heading for development in 1982, a group of Hawaiian *kūpuna* (elders) got together and created a new vision: a land for farming, where people grew crops to support their *'ohana* (families); an upland forest where folks

Photo courtesy of Waipā Foundation

came to gather medicinal plants for their own healing; a valley where streams flowed uninterrupted from mountain to sea; and a place where the *'ōlelo* (language) of Hawai'i was spoken freely as *keiki* (children) learned their native culture.

Today, the Waipā Foundation grows *kalo* (taro) in the traditional way and pounds its corm into *poi*, supplying over fifty Kaua'i families with *poi* each week. They produce organic vegetables for sale at their weekly farmers' market, and to nourish participants of their youth programs and their families, and volunteers.

They're also reintroducing native plants to the landscape, and they're hard at work restoring an ancient Hawaiian fishpond to provide a healthier estuarine environment (a semi-enclosed area of water where the river meets the sea). At their outdoor learning center, they host hands-on cultural programs for youth and other community members.

The restoration of Waipā is both environmental and cultural, and it's strengthening the link between the two. Some would call it a vision come true.

Volunteer Activities: *Lo'i kalo* (taro patch) work, organic gardening, reforestation, coastal native plant restoration, fishpond restoration, *poi* making.

When: Most activities year round. (Campsites may be available for groups between September and May—contact for details and fees.) *Poi* making, every Thursday morning.

Who: Groups preferred, high school age or older. Individuals or couples can join in *poi* making, or may be accommodated in other activities with advance notice and a flexible schedule.

Hardiness Level (5 being most difficult): 3 to 4

Advance Notice: 1 to 2 weeks.

Education: Programs are available for groups in above activities, in *kalo* and *poi* production, and in *ahupua'a* (ancient land division) resource management. (Contact for details and fees.)

Donations: Monetary donations are appreciated and tax deductible. Contact regarding equipment and supplies needs.

Contact: E-mail or phone.
Stacy Sproat-Beck or Lea Weldon
Waipā Foundation
P.O. Box 1189
Hanalei, HI 96714
(808) 826-9969
s_sproat@hotmail.com
www.waipafoundation.org

MAUI & MOLOKA'I

Community Work Day Program

Community Work Day Program (CWD) is about helping communities help themselves. As the state leader for Keep America Beautiful and a partner of local business, government, and community, their strength is providing resources wherever needed.

Sometimes help can be as simple as a cleanup—doable by anyone, at anytime. CWD will supply the gloves and bags, and offer advice and training. And they come from experience. Five times a year, they coordinate a major cleanup at Maui County's parks, shorelines, and roadways, involving approximately five thousand annual volunteers.

On a recent three-phase effort along the curvaceous Hāna Highway, volunteers swept the roadside clear of seven hundred thousand pounds of appliances and scrap metal, and over 250 abandoned vehicles. "When it looks like someone's taking care of a place, people are less likely to dump there," program director Rhiannon Chandler emphasizes. "It's a really positive cycle."

One of their ongoing projects is the greening and restoration of Kanahā Beach, a popular windsurfing spot on Maui's north shore. Led by Mike Perry, the effort to replace invasive species with native plants and trees is a dedicated labor of love. CWD also enlists volunteers to work in their native plant nursery—helping seeds grow into seedlings for distribution throughout the county—and to help shape wire into recycling bins.

If you want to join any of these ongoing efforts or are self-motivated to spearhead one of your own, CWD are the go-to people to help make it happen.

(Their reach spans the islands of Maui, Moloka'i, Lāna'i, and Kaho'olawe—contact the headquarters on Maui for details on neighbor-island projects. *See Multi-Island section, under Keep America Beautiful, for more about the national organization.*)

Volunteer Activities: Major community cleanups, self-directed cleanups, Kanahā Beach coastal restoration, nursery work, recycling bin construction.

When: Major cleanups—about 5 per year, for 4 hours. Self-directed cleanups—anytime. Coastal restoration, nursery work, recycling bin construction—contact for available dates.

Who: Both individuals and groups welcome, all ages.

Hardiness Level (5 being most difficult): 1 to 5

Advance Notice: None needed.

Donations: Monetary donations are appreciated and tax deductible. Contact regarding equipment and supplies needs. Ask about adopting a park, beach access, roadside, reef, historic site, cemetery, or other site.

Contact: E-mail or phone.
Rhiannon Chandler
Community Work Day Program
P.O. Box 757
Puʻunēnē, HI 96784
(808) 877-2524
cwdkhb@pixi.com
www.cwdhawaii.org

"It's very satisfying to work with your hands toward a meaningful goal. We feel it is important to always remember that when you pick up trash, for instance, to not leave your own *ʻōpala*, or bad spirit, behind. To be lucky enough to provide a service that many people appreciate is very gratifying. To be able to be a part of a dune restoration or to band native seabird chicks *(ʻuaʻu kani)* is incredible—we are very fortunate. . . .

"When people see you working, they generously heap on the praise and thanks—what's not to like about that? We also find that we can play a much bigger role than we ever expected. We feel that our help is really needed and appreciated.

"Volunteerism brings meaning and relevance into your life. It is gratifying to know that what you do is appreciated, but it is even more exciting when you feel that you can include others in that same experience. . . . Some people are fortunate to have a lot of discretionary time they can devote to volunteer efforts, but if we all just do what we are able to, think of the possibilities."
—Lis Richardson
Kīhei, Maui

East Maui Animal Refuge
(a.k.a. The Boo Boo Zoo)

Sylvan and Suzie Schwab came into the animal refuge business in an uncommon way—Suzie was suffering from cancer, and Sylvan saw how caring for an injured bird helped give her strength, healing strength. He brought her more injured animals.

Thirty years and many thousands of creatures later, Suzie's recovery program has become known as the East Maui Animal Refuge, or more aptly, the Boo Boo Zoo. Suzie's cancer has disappeared.

The Schwabs welcome all kinds of animals in need—on Maui, that often means deer, pigs, goats, owls, pheasants, and countless birds, cats, and the occasional dog. Dr. Ronald Moyer, an Upcountry veterinarian, provides small-animal services free of charge. Livestock vets are paid through donations. The rest of the day-to-day work is handled by Sylvan and Suzie and a handful of volunteers.

Most volunteer projects at the refuge involve caring for the animals and keeping their sanctuary pristine. On occasion, you may even help release a rehabilitated animal back to its natural habitat. As volunteer Michele

"As I push open the door to the old aviary every Thursday morning, I never quite know what to expect. Many times new injured birds are in cages, some with bandaged wings. Often young birds from Sylvan's 'nursery' upstairs have graduated to adolescent status downstairs.

"Our out-of-cage resident mynahs are very curious and clever. They swoop around, picking up pieces of lint from the dryer or straw from the walls. Many are tame, landing on a shoulder or head for fun. . . . Two very crippled hens are always somewhere on the floor. You have to watch carefully not to step on them. They love the 'old,' uneaten worms that I discard from the caged mynahs' cups, and papayas that the mynahs have rejected. . . .

"Did I mention the ibises, cardinals, pheasants, and outdoor peacocks and peahens? Cats? Ducks? Turkeys? Goats? Deer? Sheep? The list goes on and on."

—Lynn Austin
Pā'ia, Maui

"I work and interact with dozens of people every day, and don't have a lot of quiet time to think or let my mind just wander. My time at EMAR gives me that break. Being with, caring for, the animals provides balance. It's an animal refuge, but it's my refuge, too. I get back more than I give."

—Michele Chouteau McLean
Pā'ia, Maui

Chouteau McLean describes it: "The most rewarding experience in working with wildlife has to be releasing a rehabilitated animal back to the wild. . . . When you take a rehabilitated bird back to its natural habitat, there is always a few seconds' hesitation . . . before it bolts out of your arms or out of the carrier, into the sky or the woods, back to its wild life. There is nothing like it."

Volunteer Activities: Animal care (feeding, cleaning, walking, playing with, and loving the animals); upkeep, including maintenance and carpentry, of the refuge grounds.

When: Year round.

Who: Both individuals and groups welcome. Kids under 13 must be supervised by an adult. (Animals are not caged, so volunteers enter at their own risk.)

Hardiness Level (5 being most difficult): 1 to 5

Advance Notice: Contact for an appointment.

Education: Tours given by appointment.

Donations: Monetary donations are appreciated and tax deductible. Contact regarding equipment and supplies needs. Ask about the monthly Sponsor-an-Animal program.

Contact: Phone preferred.
Sylvan Schwab
East Maui Animal Refuge
(a.k.a. The Boo Boo Zoo)
25 Maluaina Place
Haʻikū, HI 96708
(808) 572-8308
thebooboozoo@gmail.com
www.booboozoo.org

East Maui Animal Refuge

Haleakalā National Park

In the short two hours it takes you to drive to the summit of Haleakalā Volcano, you pass through as many ecological zones as you would on a road trip from Mexico to Canada. There are cinder cones in crimson, burnt orange, and ochre; there are spiky silverswords that burst forth in a once-in-a-lifetime bouquet before the entire plant dies. You might encounter a rare gaggle of endangered *nēnē* (Hawaiian geese and the state bird), or find yourself wandering through a misty cloud forest lush with native ferns.

Most volunteer projects land you somewhere in this otherworldly landscape, helping native species survive against invasive plants and animals, such as feral goats and pigs. There are many opportunities available for both individuals and groups, plus short service trips offered by the Friends of Haleakalā (listing follows).

Then there is Kīpahulu. This coastal jewel lies on Haleakalā's verdant backside, about ten miles past remote and rural Hāna town. Its 'Ohe'o Gulch is a popular destination with adventure seekers there to explore the area's sparkling pools, or to trek through dense bamboo to see waterfalls free-falling from sheer cliffs overhead. Rare native vegetation thrives here in the caring hands of park staff and dedicated volunteers.

Volunteer Activities: Within the crater—weeding, trail maintenance, visitors' center assistance (see www.nps.gov/hale/supportyourpark/volunteer-opportunities.htm for descriptions and more opportunities). In Kīpahulu area—maintaining native plantings.

When: Year round, project lengths vary greatly. (See Web site or contact.)

Who: Individuals welcome for longer-term projects (see Web site for various age requirements) or in Kīpahulu area (all ages welcome). Groups preferred for short-term service trips; kids must be supervised by an adult.

Hardiness Level (5 being most difficult): Weeding, Kīpahulu area campground maintenance, visitors' center assistance—1. Trail maintenance—3 to 5. Group service trips tailored to group level.

Advance Notice: Individuals—see Web site; groups—at least 1 month.

Education: Cultural activities, presentations, short hikes, and Junior Ranger activities take place daily. Biweekly hikes are given through the Waikamoi preserve (require 1-week advance notice). Evening programs are offered as announced. Contact the education coordinator at hale_interpretation@nps.gov, (808) 572-4453, for details or regarding school group visits. All programs are free.

Donations: Funds for volunteer and educational programs accepted at donation boxes in visitors' centers. Only donations made through Friends of Haleakalā National Park are tax deductible (see following listing).

Friends of Haleakalā National Park—Matt Wordeman

Contact: E-mail or phone.
Volunteer Coordinator
Haleakalā National Park
P.O. Box 369
Makawao, HI 96768
(808) 572-4487
HALE_VIP_Coordinator@nps.gov
www.nps.gov/hale/supportyourpark/volunteer.htm

Friends of Haleakalā National Park—Matt Wordeman

Friends of Haleakalā National Park

The Friends of Haleakalā National Park are just that—people who care about this rare and stunning landscape as much as the folks hired to protect it. They're particularly fond of service trips—hiking into Haleakalā Crater for a few days, pulling weeds, blazing trails, and relaxing in one of three rustic wilderness bunkhouses come nightfall. The trips happen once a month and allow time for volunteers to explore the remarkable scenery each day.

Volunteer Activities: Invasive species removal and planting native species.

When: Monthly, for 2 to 4 days.

Who: Both individuals and groups welcome, age 7 and older.

"We had an incredible experience on our service trip with Friends of Haleakalā [FoH]. We went as a family of four, with our two teenagers (a high school and a college student). We were awed by the beauty of the crater, and hiking all the way across to Palikū allowed us to see the diversity of Haleakalā. I was impressed with the dedication of the FoH volunteers. The education they provided about indigenous plants and invasive species was fascinating, even for a person like myself who is not passionate about plants! . . .

"I'd say the best thing for me was seeing my teenagers involved, interested, and learning about conservation and stewardship of these lands we are fortunate to share. We learned about the indigenous plants, as well as some natural history of Haleakalā and Hawaiian history, and the stargazing was phenomenal. . . .

"I had always wanted to hike in Haleakalā Crater, and this was a great way to do it, with great support and education."

—Kathleen Elliott, PA-C, and David Jones, MD Honolulu, O'ahu

"It surprises me how many different types of people are interested in the service trips. You'll get families up there, and people who live in tents in the middle of the woods, people in school. The first time there were these two eighty-year-old women in the group—they did the whole ten-mile hike and carried their own packs! It was pretty impressive. You meet a lot of interesting people up there, but not necessarily the people you think you're going to meet.

"It's rewarding to know you're doing something productive and helping your environment . . . and you're up there in this amazing scenery where not many people go. A lot of people go to the summit, but not down into the crater, where you're miles from anywhere else."
—Shelley Harmer
Lahaina, Maui

Hardiness Level (5 being most difficult): 2 to 4, depending on which cabin is the destination (one-way hikes from trailhead to cabins range from about 4 to 10 miles).

Advance Notice: As much as possible, up to a few days ahead if openings are available.

Donations: Monetary donations are appreciated and tax deductible. Ask about the Adopt-a-Nēnē (endangered Hawaiian goose) program.

Contact: E-mail or phone.
Friends of Haleakalā National Park
P.O. Box 322
Makawao, HI 96768
Farley Jacob, (808) 248-7660
Matt Wordeman, mrword@hawaii.rr.com, (808) 876-1673
www.fhnp.org

Friends of Haleakalā National Park—Matt Wordeman

Hawai'i Nature Center—Maui

In the Hawai'i Nature Center at the base of majestic 'Iao Valley, kids are studying bee antennae through a microscope. Others are lying facedown, arms outstretched like dragonflies, on a virtual-reality flight along the 'Iao Stream. Some kids are out traipsing through the rain forest, studying how the ecosystem works in a dirt-under-the-fingernails sort of way.

On this thirty-five-acre parcel of nature's playground, help is always needed. That might mean removing invasives and replanting native species, clearing the jungle away from ancient *lo'i kalo* (taro patches), painting the facility's buildings—basically, anything you have time to do. "We don't always have the personnel to do everything we need to do," says volunteer coordinator Jay Franey, "so well-focused, well-intended volunteers often save the day in our daily operations."

What better way to save the day than by spending it in nature?

(See Multi-Island section, under Hawai'i Nature Center, for more about the organization itself.)

Volunteer Activities: Removing invasive species, planting natives, clearing ancient terraces, landscaping, painting.

When: Year round.

Who: Both individuals and groups welcome, age 12 and older.

Hardiness Level (5 being most difficult): 4

Advance Notice: 1 week.

Education: Interactive Nature Museum, guided Rainforest Walk, extensive and ongoing kids' programs. (See Web site for details and fees.)

Donations: Monetary donations are appreciated and tax deductible.

Contact: E-mail preferred.
Jay Franey
Hawai'i Nature Center—Maui
875 'Īao Valley Road
Wailuku, HI 96793
(808) 244-6500 or (888) 244-6503
jay@hawaiinaturecenter.org
www.hawaiinaturecenter.org

Hawai'i Wildlife Fund—Maui

Sleeping on the beach has its lures—black night skies, sand in your hair, the lullaby of waves tumbling at the foot of your bed. Then the alarm goes off and you're up again. It's 2:00 a.m. You scan the sand for turtle tracks. Nothing. You look for baby turtles wandering in the darkness, instinctively trying to reach their saltwater home. Not yet. You reset the alarm and lean back against a cushion of sand, breathing in the seaweed air, knowing that if just one in a hundred hatchlings survives its journey from nest to ocean tonight, you'll have done your job.

For those of you more prone to sun seeking than night vision, you can join the Hawai'i Wildlife Fund (HWF) response team that watches over Hawaiian monk seals when they haul out onto Maui's beaches. You'll take turns guarding a taped-off zone around the lounging seal, while educating onlookers about these glorious creatures and their behaviors.

On regular occasions, the HWF also takes to the shores with gloves and trash bags, picking up litter where it doesn't belong. And they run Makai Watch, where you can help trained naturalists spread the good word of respecting the marine environment.

(See Multi-Island section, under Hawai'i Wildlife Fund, for more about the organization itself.)

Volunteer Activities: Hawksbill sea turtle nest watch (plus dune habitat restoration, fence mending, beach cleanups), monk seal watch, Makai Watch.

When: Most activities—year round, for 2- to 5-hour shifts. Hawksbill nest watch—summer only, for 2- to 6-hour nighttime shifts.

Who: Both individuals and groups welcome. Small children can participate in most programs if supervised by an adult.

Hardiness Level (5 being most difficult): Hawksbill nest watch—1 to 3. Monk seal watch—1. Makai Watch—2.

Advance Notice: Hawksbill nest watch—at least 1 week. Monk seal watch—none needed. Makai Watch—1 week preferred.

Education: In addition to free Makai Watch education, they offer 6-week naturalist training courses through Maui Community College (contact for details and fees).

Donations: Monetary donations are appreciated and tax deductible. Contact regarding equipment and supplies needs. Ask about the Adopt-a-Whale/Dolphin/Monk Seal/Turtle/Coral Reef programs.

Contact: E-mail preferred.
Hannah Bernard
Hawai'i Wildlife Fund
P.O. Box 790637
Pā'ia, HI 96779
(808) 280-8124
wild@aloha.net
www.wildhawaii.org

"It's really fun. You sleep on the beach; you get up throughout the night at intervals to either check for the female turtle coming onto the beach to lay her nest, or you check a nest to see if there are any tracks, if there are baby turtles that have hatched and are walking around. It's really an amazing thing to see baby sea turtles climbing out of the sand—the sand starts to almost boil, like boiling water. When they come out their flippers are so long compared to their bodies; they look like gangly horses.

"You have to have a lot of patience. Sea turtles have about a two-week window of hatching time, so if you really want to see baby turtles, you have to camp every night for at least a few weeks, or the entire season, several months.

"The poor hawksbills are so endangered, but they don't get any press and yet they're the most beautiful. It's a very cool opportunity to see a really endangered species.

"I'm a student at the University of Oregon. It has not cost me any more than a term at school regularly costs and has been a thousand times more informative. Real-world experience is worth way more than classroom lectures!"

—Abigail DeYoung
Eugene, Oregon

"Even though I am just one person, if enough of us 'one persons' band together we can make an even bigger difference. Why else would some of us give up our only morning to sleep late to walk the beach looking for turtle tracks of nesting turtles? That is how my volunteer experience began.

"I can only walk one morning each week because I work full time at the Maui Ocean Center and also paddle canoe before I go to work. The second summer I volunteered, I could still only walk one morning a week, but I was sleeping more nights on the beach than I was at home. . . .

"To see a baby hawksbill turtle, whose whole body—legs and all—would fit into the palm of your hand, make its way to the ocean is one of the most joyous sights you will ever experience. When Hawai'i had its 6.8 earthquake [October 2006], Cheryl and I were sitting on the beach waiting for more turtles to emerge from the nest—did we think about saving ourselves and running for higher ground in case there was a tsunami? Of course not—we wanted to see if the quake would cause more turtles to hatch. . . .

"I did not know what to expect when I first started volunteering—it turned out to be such an opportunity not only to help the turtles and the monk seals, but to meet other volunteers with a similar interest. I feel so blessed to be associated with Nicole [Davis], Hannah [Bernard], and especially Cheryl [King, of the HWF team]. The world is a better place because of them, and I am a better person for knowing them."

—Norma Clothier
Kīhei, Maui

Honokōhau Valley Project

When Job Cabato returned to Honokōhau Valley, the valley of his ancestors, he thought it was to master slack-key guitar—he didn't plan on becoming a master of taro farming. His mentor, slack-key guitar legend George Kahumoku, feels the two are inseparable: "It's one and the same to me, just a different art form. One is food for the soul, the other food for the stomach."

Job agrees. Working in the soil, he adds, takes the same level of dedication, plus it strengthens the fingers for better playing. As an example, the dense jungle around his home was grown from seed by his own hand, with the *kalo* (taro) as its centerpiece. Job plants taro in both the *māla* (dryland) and *loʻi* (wetland) styles. "Taro honors our *kūpuna* [ancestors]," Job says. "We perpetuate culture when we keep the taro alive and growing."

With many different kinds of taro on the land, there's always plenty of work—more work than one man can do alone. That's why Job started volunteer days on Sundays, when anyone can join him in this remote West Maui Mountains locale to learn in the traditional way—by doing.

Kirsten Whatley

Job hopes to bring taro back to the land in numbers that can sustain a whole valley, as in olden times. It's going to take an ongoing effort, but he's committed. In his words, "If you work hard at what you do, it's beautiful. Like the music—if you play good, it's beautiful."

Volunteer Activities: *Lo'i kalo* (taro patch) restoration.

When: Every Sunday morning, for 3 hours. (Camping may be available—contact for details.)

Who: Both individuals and groups welcome, all ages.

Hardiness Level (5 being most difficult): 3

Advance Notice: 1 week preferred.

Education: Job is an accomplished slack-key guitar player and loves to share his music. (Contact him for details.)

Donations: Monetary donations are appreciated—signify the donation is for taro patch restoration. (Ask about tax deductibility.)

Contact: Phone preferred.
Job Cabato
Honokōhau Valley Project*
P.O. Box 10246
Lahaina, HI 96761
(808) 665-0628
* The taro patch is located in Honokōhau Valley, about 1 hour's drive from Wailuku. (Call Job for directions.)

"You can't find the Honokōhau Valley on most maps. Only about a hundred people live down its single dirt lane within dense vegetation, and the only way out is to go back the way you came. . . .

"I spent a summer on the land here, learning as much as I could from Job. Being a landscape designer, I also designed for him an herbal healing garden and coconut road for future visitors to enjoy. We planted seeds and propagated tropical plants, and he taught me about taro farming—taro is the essential Hawaiian staple food. There are certain procedures to follow in the taro patch, like which direction to plant the corm for the greatest energy.

"Learning in the old way from Job made me realize how important the traditions of Old Hawai'i are, and how quickly they're being lost."
—Jonathan Silverman
San Francisco, California

"It's important for a community to take charge of their own destiny. I was just at a community meeting telling people, instead of planting grass, plant something you can eat! . . . We can't be dependent on Costco or Safeway, or whatever comes from outside. We've got to change our mind-set. We've got to think of seven generations, forward and back."
—George Kahumoku
Lahaina and Kahakuloa, Maui

Keālia Pond National Wildlife Refuge

Nearly seven hundred acres of wetlands in south-central Maui serve as gathering grounds for feathered family reunions. At Keālia Pond each winter, the endangered *ae'o* (Hawaiian stilt) and *'alae ke'oke'o* (Hawaiian coot) host migratory shorebirds, which crowd the edge of the pond to feast on aquatic insects. The birds are also joined by migrating waterfowl, which make the pond their Hawaiian getaway until temperatures elsewhere rise the following spring.

Open water, shallow mudflats, and pockets of vegetation form a soothing backdrop to volunteering at the pond. You might help with propagation and planting of native species, refuge maintenance, or overall wetlands restoration projects.

For volunteers who can make a longer-term commitment, you might be called on for turtle patrol, or Dawn Patrol, as it's known. South Maui beaches provide nesting sands for hawksbill and green sea turtles, and volunteers are crucial in identifying signs of nesting activity. This helps biologists in their monitoring efforts and in ensuring beach habitats stay protected.

On Dawn Patrol, volunteers go out first thing in the morning, before human footprints make scouting for turtle tracks impossible. Volunteers Ed

U.S. Fish and Wildlife Service

"An ecology teacher asked my class to close our eyes and imagine 'nature.' After a couple minutes, each of us described our thoughts. Only myself and another described nature as us (people) being part of it. All other classmates talked about the forest, the ocean, the animals, trees, etc.

"This was an eye-opener for us all in that almost nobody considered ourselves part of nature. . . . I make it a point to proactively include myself with nature, and [U.S.] Fish and Wildlife [Service] provides a fulfilling outlet. Discovery Channel is great, but it's not the real thing—you can't hear the live sounds, smell the air, feel the temperature nor experience the collective whole. . . .

"Living on Maui has reminded me how people used to be one with the land in all things. Every day, I learn more examples [of] how the Hawaiians revered the land, sea, sky, and themselves as a single and universal interaction—ask any Hawaiian and they will tell you their eldest brother is the taro plant—there is a reason for it!

"As we increase our pace concerning economic growth combined with self-conveniences, we have forgone much of our stewardship responsibilities, and we owe it to ourselves to take a step back and include nature as an integral part of our daily existence."

—Steve O'Toole
Lahaina, Maui

"Hawai'i is such a beautiful place to be. I was amazed and disappointed to hear that so many plants I saw in the city were not even native to Hawai'i.

"My family and I visited Maui last year and we asked to help at Keālia Pond National Wildlife Refuge for a morning. We learned a lot about native plants and made some new cuttings for the nursery and planted almost one whole area in the wetlands. These were for the endangered stilt and coot that live in the wetlands, so the time was really valuable, for us and the birds. It's peaceful to work there and the birds just kept eating close to us.

"Even though we were on vacation I really felt like we helped the birds and the people working there. They have a lot more to do."

—Sarah Stencil
Madison, Wisconsin

and Loretta Klimczak describe it like this: "So many people ask us why we dig up the nest two days after the turtle emergence. More times than not we have saved little turtles that probably would not have made it on their own. Hawai'i gets tourists from all over the world, and at one of our digs a couple from France remarked that the turtle excavation was the highlight of their vacation. This is what makes it special."

(See Multi-Island section, under Hawaiian Islands National Wildlife Refuges, for more about the agency itself.)

Volunteer Activities: Wetlands restoration work—propagating and planting native species, maintenance, wetlands restoration; Dawn Patrol.

When: Wetlands restoration work—year round, for 2 or more hours per project. Dawn Patrol—June to September, 2-month minimum commitment, for daytime shifts (free volunteer training usually occurs in May, but can be arranged at other times).

Who: Wetlands restoration work—both individuals and groups welcome. (Kids under 18 must be supervised by an adult.) Dawn Patrol—individuals or families, age 18 and older preferred.

Hardiness Level (5 being most difficult): Wetlands restoration work—1 to 5. Dawn Patrol—1 to 3.

Advance Notice: At least 2 weeks.

Education: Tours and environmental education given by appointment, with at least 1-week advance notice.

Donations: Contact regarding equipment and supplies needs.

Contact: E-mail or phone.
Glynnis Nakai
Maui National Wildlife Refuge Complex
P.O. Box 1042
Kīhei, HI 96753
(808) 875-1582
glynnis_nakai@fws.gov
www.fws.gov/pacificislands/wnwr/mkealianwr.html

Kīpahulu ʻOhana
Kapahu Living Farm

In the fertile valley of Kīpahulu, tucked within the folds of Haleakalā's jungled east side, the ancient Hawaiian practice of taro cultivation is thriving.

Meet John and Tweetie Lind, the devoted stewards of 2 1/2 acres of *loʻi kalo* (taro patches), lovingly restored by the Linds to perpetuate the traditional ways of their Hawaiian ancestors. Started in 1995, the Kīpahulu ʻOhana aims to teach others about the "ways of old" through demonstrations and hands-on activities.

Volunteering at Kapahu Living Farm means getting your hands—even your feet—in the mud of the *loʻi*, where you'll work side by side with farmers whose families have lived here for generations. You might also learn to pound *poi*, the cherished result of the taro harvest, or prepare *lau hala*, the leaves of the *hala* tree, for weaving into traditional crafts.

There are no set fees for the ʻOhana's service-based learning when your time is dedicated to the shared hands-on work, but donations are greatly encouraged.

Scott Crawford, Kīpahulu ʻOhana

Volunteer Activities: Taro cultivation and cultural activities.

When: Year round, from 2 hours to several days.

Who: Groups only, age 4 and older.

Hardiness Level (5 being most difficult): 2

Advance Notice: 1 week.

Education: Interpretive hikes of either 2 hours or 3 1/2 hours offered from Haleakalā National Park (Kīpahulu area) to Kapahu Living Farm. Individuals welcome, age 8 and older, or infants that can be carried. (Minimum of 2 people. See Web site for details and fees.)

Donations: Monetary donations are appreciated and tax deductible. Contact regarding equipment and supplies needs.

Contact: Web site, e-mail, or phone.
Kīpahulu 'Ohana
P.O. Box 454
Hāna, HI 96713
(808) 248-8558
tours@kipahulu.org
www.kipahulu.org

Scott Crawford, Kīpahulu 'Ohana

Maui Coastal Land Trust
Waihe'e Coastal Dunes and Wetlands Refuge

It was supposed to be a golf course—277 acres of coastal wetlands and sand dunes that were once an ancient fishing village, where *heiau* (traditional places of worship) and extensive burial sites still exist. But the Maui Coastal Land Trust had other visions.

They wanted to ensure the *heiau* and burial sites remained intact. They dreamed of the inland fishpond and *lo'i kalo* (taro patches) becoming demonstration plots for cultural agriculture. They envisioned restoring the habitat for endangered shoreline birds such as the *ae'o* (Hawaiian stilt) and *'alae ke'o ke'o* (Hawaiian coot). They knew their visions were going to take a lot of helping hands.

Every Friday morning, a team of volunteers descends upon the site with gloves, tools, and determination. The group focuses mainly on

"We thought volunteering would be a fun way to see a less-visited part of Maui and meet local people not involved in the tourist industry. It also provided some exercise to work off the effects of all the tasty food we'd eaten. . . .

"There seemed to be a number of options for activities, depending on your energy level that day. Some people dragged branches, others raked up all the debris left behind, others dug up roots using pickaxes. . . .

"Our organizer, Scott, invited the volunteers to join him and the students at the end of the morning when he gave them a guided tour of the site and a brief talk about the history and ecology of the site. . . . It was not so much learning something [for us], but just getting outdoors with local people and doing something useful. It makes the holiday more memorable than just doing the usual touristy things."

—John and Heather Wingert
Burnaby, British Columbia,
Canada

Maui Coastal Land Trust

clearing away invasives to make room for the native plants that once flourished here, such as *'ahu'awa* (a local sedge), *kaluhā* (papyrus), and *'ākulikuli* (a coastal succulent). By lunchtime, the Waihe'e coast is that much closer to regaining the diversity and grandeur of its former self.

Volunteer Activities: Wetlands restoration—removing invasive plants and replanting native vegetation.

When: Every Friday morning, for 4 hours.

Who: Both individuals and groups welcome, all ages. (For larger groups, a donation is requested to support additional staff.)

Hardiness Level (5 being most difficult): 3

Advance Notice: Individuals—1 day; longer preferred for groups.

Education: The first Saturday of each month, they offer a free guided 2 1/2-mile walk through the refuge. Advance notice of 1 to 2 weeks recommended—only 25 spaces available.

Donations: Monetary donations are appreciated and tax deductible. Contact regarding equipment and supplies needs.

Contact: Phone preferred.
Scott Fisher
Maui Coastal Land Trust
P.O. Box 965
Wailuku, HI 96793
(808) 244-5263
info@mauicoastallandtrust.org
www.mauicoastallandtrust.org

"I get more appreciation for a place by working on it. I think it's more worthwhile to get dirty—because when you touch a place, you remember it.

"We really want the kids to get back to their roots, hands on, and know where their ancestors came from, so that they're proud of who they are. So that they have more respect."

—Laua'e Murphy
Kamehameha Schools
Enrichment Department
Kahalu'u, O'ahu

Maui Cultural Lands

Projects Malama Honokowai and Malama Hanaula

Beyond the grand resorts of Kā'anapali, beyond the paved sidewalks and asphalt lots, a village is being reborn in a fold of the West Maui Mountains. It's not the first time the Honokōwai Valley has seen people. In fact, in the time of kings, it was the breadbasket of this important region.

Ed Lindsey thinks the valley can return to a model of agricultural and cultural significance. So does his wife, Puanani, his family, and his cohort in conservation Rene Sylva. Together, they've identified three miles of archaeological sites beneath a thick layer of invasive plant growth. Every Saturday, Ed takes volunteers into the valley to help chip away at this overgrowth, to replant native species and stabilize the archaeological sites—all the while learning by doing. He hopes Project Malama Honokowai will set a precedent for preserving and protecting lands up and down the coast.

Ed works with developers too, figuring that by sharing the land's cultural importance, more of it can be saved from the bulldozer before it's too late. "As my four-year-old grandson asked me, 'Grandpa, why is this important?'

"We don't know the secrets that these stones hold. But we do know that if these stones are gone, those secrets are gone forever and that story is gone forever. And something that would enrich the present and future life of all those who live on Maui, whether they're Hawaiian or not, is gone forever. . . .

"What's happening in Honokōwai Valley is a miracle. And it's really, really going to be a gift for future generations."

—Lucienne de Naie
Huelo, Maui

"Being five years on this particular project, when the little seedlings become trees, and the trees are over your head, and the bushes become gigantic shelters for the understory, you start looking at these things and you go, 'Wow . . . we've actually made an impact.'"

—David Metz
Ha'ikū, Maui

> "I think every child who lives in the islands should experience something like this. They can really learn and have hands-on [experience] of what *mālama 'āina* [caretaking the land] really means. . . . These are things that they're learning in the classroom every day, but they never get to touch, they never get to feel, they never get to see. This just puts all the pieces together for them. . . .
>
> "They'll be the ones to carry on the legacy of our *kūpuna* [elders] and our ancestors."
>
> —Maile Ramos
> Teacher,
> Princess Nahi'ena'ena
> School
> Lahaina, Maui

> "I think people come to Hawai'i and there's less and less of Old Hawai'i for them to see. I treasure being part of a project that makes Old Hawai'i available for people to see in the future. Everything seems to not exist outside the valley when you're in here."
>
> —Donna Kroetsch
> Launiupoko, Maui

. . . My answer was, 'Because that's who we are,'" says Ed. "If you destroy these places, who are we?"

Another of these places is the windswept territory near Kaheawa Wind Power, a wind farm above Mā'alaea, just north of Kīhei. That's where you'll find Ed on Sundays, leading Project Malama Hanaula, a reforestation effort to return this south Maui landscape to its thriving original state. In Ed's eyes, it's a responsibility to the land that can't be ignored. *Mālama ka 'āina* (take care of the land), he says, and it will give back more than you put in.

"That's the main effort of this place—to learn, rediscover, to heal and to form partnerships, and to form extended families. That's the Hawaiian style."

Volunteer Activities: Archaeological stabilization, removing invasive species, reforestation.

When: Project Malama Honokowai—every Saturday, for 6 hours. Project Malama Hanaula— every Sunday, for 6 hours.

Who: Individuals and small groups welcome, all ages. Youngsters must be supervised by an adult.

Hardiness Level (5 being most difficult): 1 to 5

Advance Notice: 1 day (to let him know number of people in your party).

Donations: Monetary donations are appreciated and tax deductible.

Contact: Phone preferred.
Ed Lindsey
Maui Cultural Lands
1087-A Po'okela Road
Makawao, HI 96768
(808) 572-8085
www.mauiculturallands.org

Maui Nui Botanical Gardens

What's the best way to preserve and protect native species? Grow 'em. That's what they do at Maui Nui Botanical Gardens, on five acres of gorgeously landscaped land smack in the middle of busy Kahului. The collection includes plants that are endemic (found only in Hawai'i) and indigenous (found in

Kirsten Whatley

Hawai'i as well as other places, arriving here without the help of man), growing alongside the plants carried here by early Polynesian settlers.

Established by Rene Sylva in 1976, the gardens feature plants naturally adapted to Maui Nui's* unique environments, many from seeds hand collected in the wild by Sylva himself. But look closer, behind the foliage, and you'll discover they're not just growing plants here, they're raising awareness of a living culture. Workshops include making *kapa* (bark cloth), traditional cordage, and plant dyes. Then there are Arbor Day giveaways—where thousands of native trees are handed out free to good homes—and their popular native plant sales.

To volunteer, you must first become a member (forms and fees are on their Web site). This rewards you with updates on the native plant world through their quarterly newsletter, and invites you to the weekly Weed and Pot Club, an ongoing volunteer work party designed to keep the gardens thriving.

* *Maui Nui includes the islands of Kaho'olawe, Lāna'i, Maui, and Moloka'i.*

Volunteer Activities: Garden work, nursery work, occasional special projects (building compost piles, working with sugarcane, demolition of small structures).

"When I moved to Maui, I wanted to learn not only how to grow things but how native people here use native plants. If I just went up the mountain myself, I would never notice as much as I do at the concentrated location of the gardens. It's really seeing a whole variety of plants so rare you wouldn't see them otherwise.

"I've learned there are some very committed people here on Maui restoring native plants, and educating the public about the traditional uses of the plants. It's very inspiring; they're really dedicated. It's nice to be around them."

—Anudeva Stevens
Huelo, Maui

"I've volunteered my whole adult life, and I've had all kinds of volunteer experiences, from the worst to the best. This is the best! The place itself creates a warmth and good feeling. They make you feel like part of the extended family.

"I do it because I love plants and I wanted to learn about Hawaiian plants. And if I can do something good while I'm learning, that works for me."

—Renee Leiter
Wailea, Maui

When: Every Wednesday morning, for 2 hours.

Who: Both individuals and groups welcome; individuals preferred. Kids must be supervised by an adult.

Hardiness Level (5 being most difficult): Garden and nursery work—2. Special projects—3 to 5.

Advance Notice: Individuals—1 week; groups—1 month.

Education: Public classes include native Hawaiian plant–related crafts, plant identification, propagation, and pruning. (Contact for details and fees.) Tours and special educational programs available by appointment.

Donations: Monetary donations are appreciated and tax deductible. Contact regarding equipment and supplies needs. Ask about the Adopt-a-Bed program.

Contact: Phone preferred.
Kea Hokoana-Gormley
Maui Nui Botanical Gardens
150 Kanaloa Avenue
P.O. Box 6040
Kahului, HI 96733
(808) 249-2798
mnbg@maui.net
www.mnbg.org

Maui Restoration Group

The *koa* stands as the largest tree in the native Hawaiian forest. In Hawaiian, its name means "brave" and "hero." Its wood can be turned into canoes, surfboards, furniture, and *'ukulele*. The fact that its populations have been decimated on the island is a disheartening symbol for the fate of all Hawai'i's forests.

On leeward Haleakalā, less than 10 percent of the original forests remain. Invasive European and African grasses are encroaching upon what's left of the understories—the shrubs, ferns, and mosses that shelter the forest floor. Since 2003, the Maui Restoration Group and the motivating force behind it, research biologist Art Medeiros, have been trying to reverse this demise. On more than forty-three thousand acres stretching from Makawao to Kaupō, they've pulled literally millions of weeds and planted over thirty-five thousand native seedlings.

Volunteers are the backbone of this effort. On semimonthly trips into the forest—such as at Auwahi or Pu'u-makua—they work alongside experts in ethnobotany, natural history, and land management, all the while learning as they plant trees together, remove nonnative species, and collect seeds.

"Auwahi was one of the richest forests in Hawai'i," says Art. "Then everything was gone but the trees—there was nothing under the trees. When we first came on the scene, most people thought there was no hope for it.

Erica von Allmen

They called it a museum forest. But now, not only has the forest come back, most has been done by the community of Maui—people have saved the forest.

"No one ever helps the forest as much as the forest helps them. *Aloha 'āina* [love of the land] is a concept, but how do you *aloha* the *'āina* if you don't know the *'āina*? People come out here and they leave the experience different. It's something profound."

Volunteer Activities: Removing nonnative species, planting trees, collecting seeds.

When: Twice monthly, on Saturdays, for 8 hours. (See Web site for upcoming dates.)

Who: Both individuals and groups welcome, ages 14 to 65.

Hardiness Level (5 being most difficult): 4

Advance Notice: At least 2 days.

Donations: Monetary donations are appreciated and tax deductible. Contact regarding equipment and supplies needs.

Contact: E-mail preferred.
Andrea Buckman
Maui Restoration Group
P.O. Box 652
Makawao, HI 96768
(808) 573-8989
auwahi@yahoo.com
www.lhwrp.org

Na Ala Hele

Na Ala Hele—Maui

As a volunteer with Na Ala Hele on Maui, you might find yourself hiking through the crisp air of Olinda, where a forest of native *'āla'a, hala pepe,* and *koa* trees stands tall. Or you may end up trekking along Haleakalā's spine, with cinder cones and craters at your side. Some projects might lead you on a hike above Maui's eastern shore, or upon the historic Old Lahaina Trail with views of Kaho'olawe and Lāna'i on the horizon.

While you're enjoying the scenery, there'll also be trails to maintain or maybe even construct. A small trade-off for the opportunity to experience Maui's footpaths firsthand.

(See Multi-Island section, under Na Ala Hele, for more about the agency itself.)

Volunteer Activities: Trail maintenance and construction.

When: Year round.

Who: Both individuals and groups welcome, age 13 and older. Kids must be supervised by an adult. Much of the work is done at higher elevations, which may be hazardous to those with heart conditions.

Hardiness Level (5 being most difficult): 4

Advance Notice: 1 day.

Donations: Monetary donations are appreciated and tax deductible. Contact regarding equipment and supplies needs. Ask about the Adopt-a-Trail program.

Contact: E-mail or phone.
Kevin Cooney
Na Ala Hele—Hawai'i Trail and Access System
54 South High Street, Suite 101
Wailuku, HI 96793
(808) 873-3509
Kevin.P.Cooney@hawaii.gov
www.hawaiitrails.org

Native Hawaiian Plant Society

There are just over a thousand native Hawaiian plants on the islands. There are an estimated fifteen to twenty thousand nonnative plants here. That makes the odds of native survival fifteen- or twenty-to-one. Not very promising.

And it's not just invasive plants that are pushing the natives toward extinction, but the introduced animals and insects that prey on them, and the fungi and diseases that weaken their resistance. Plants found nowhere else in the world are fighting a battle from all sides just to stay alive in their island home.

Dedicated conservationist Rene Sylva is helping lead this battle. And one of his many brainchildren is the Native Hawaiian Plant Society

Irene Neuhouse

(NHPS). Their method is to go into Maui's outback to rally around the natives—fencing out feral animals, clearing weeds, and propagating native starts in hopes that they'll flourish and multiply.

An example is Hawai'i's state flower, *ma'o hau hele*, a yellow hibiscus that is extinct on some islands, endangered on others. On more than ten thousand square feet in a remote corner of the wilderness, NHPS volunteers have erected a fence around the flowers to keep outsiders at bay. Within the "exclosure," they pull out competing weeds to allow the hibiscus to absorb nutrients while basking in the sunshine and rain.

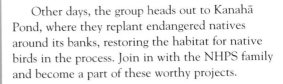

Other days, the group heads out to Kanahā Pond, where they replant endangered natives around its banks, restoring the habitat for native birds in the process. Join in with the NHPS family and become a part of these worthy projects.

Volunteer Activities: Eradicating invasive species and replanting natives.

When: Service trips—second Saturday of most months, for 6 to 8 hours. Kanahā Pond—every Thursday, all morning.

Who: Both individuals and groups welcome. Kids must be supervised by an adult and able to handle a full day out. (Service trip locations are remote and far from conveniences.)

Hardiness Level (5 being most difficult): 3

Advance Notice: A few days.

Education: Public lectures are held in February, June, and November. (See Web site for details.)

Donations: Monetary donations are appreciated and tax deductible.

Contact: E-mail preferred. For service trips, contact trip leader given on Web site. For Kanahā Pond, contact Lorna Hazen, lornajack@clearwire.com.
Native Hawaiian Plant Society
P.O. Box 5021
Kahului, HI 96733-5021
einew@hotmail.com
www.nativehawaiianplantsociety.org

"We moved to Maui almost ten years ago, and my husband found a reference to the NHPS online. We joined, I went on a service trip, and never looked back. After several service trips, I was asked if I were interested in joining the board. I have been vice president and am now secretary. NHPS provides me with a way to enjoy myself, learn about Hawai'i's unique flora (plant identification has been a hobby since childhood), and contribute to its preservation at the same time.

"Much of the remnants of Maui's native plants are difficult to find and identify. . . . Membership in the NHPS has been my major resource for learning to identify the native and alien plants on Maui."

—Irene Newhouse
Kīhei, Maui

Pacific Whale Foundation

Pacific Whale Foundation (PWF) researchers might be more at home on the water than on land. Since 1980, they've been diving into the monumental effort of saving Hawai'i's humpback whales, helping bring them back from near extinction. They also help

© Pacific Whale Foundation

save the islands' coral reefs, turtles, toothed whales, and dolphins from threats to their natural habitats. Through Pacific Whale Foundation's popular marine ecotours, they've reached nearly three million people, helping them become more responsible stewards of the ocean.

One way to learn about Pacific Whale Foundation firsthand is by volunteering at one of their annual whale events, typically held each February: Whale Day is part festival, part awareness raising, and a fun-filled celebration of these beautiful behemoths; Run for the Whales is a race on their behalf; and the Great Whale Count will teach you more than just how to count these mammals, but how to identify their behaviors such as tail slaps, blows, and breaches, a valuable resource for the whales' ongoing conservation.

The foundation also runs a Volunteering on Vacation program, where they connect visitors with many of the important projects featured in this book. (Call the foundation up and be matched with volunteer programs that suit your schedule and energy level—if you volunteer for three hours or more, you get a Volunteering on Vacation T-shirt.) In addition, Pacific Whale Foundation leaders themselves take volunteer groups to Haleakalā National Park to clear invasive pine trees at the park's seven-thousand-foot gateway, and to help in the park's native species greenhouse.

If you'd rather do good on your own time, Pacific Whale Foundation also offers free Beach Cleanup Kits. Stocked with trash bags, gloves, and helpful tips, you can make your next day at the beach a productive one. There are also ongoing opportunities at Pacific Whale Foundation's Ocean Stores or office and, for qualified volunteers, on the vessels themselves.

Volunteer Activities: Annual whale event assistance, tree clearing and greenhouse work, do-it-yourself beach cleanups, assistance in the stores or office, help on vessels (for qualified volunteers—contact for details).

When: Whale events—every February. Tree clearing and greenhouse work—every third Sunday, for 3 hours of service plus transportation time. Beach cleanups/stores or office assistance—anytime.

Who: Both individuals and groups welcome, age 10 and older.

Hardiness Level (5 being most difficult): Whale events—1 to 3. Tree clearing and greenhouse work—3 to 4. Beach cleanups/stores or office assistance—1 to 2.

Advance Notice: Whale events—1 month preferred. Tree clearing and greenhouse work—at least 2 days. Beach cleanups—at least 1 day. Stores or office assistance—2 weeks preferred.

Education: Free Making Waves lectures are given monthly. Marine ecotours, full-moon tide explorations, and camps for kids are ongoing. (See Web site for details and fees.)

Donations: Monetary donations are appreciated and tax deductible. Contact regarding equipment and supplies needs. Ask about the Adopt-a-Whale and Adopt-a-Dolphin programs.

Contact: E-mail or phone.
Community Outreach Coordinator
Pacific Whale Foundation
300 Māʻalaea Road, Suite 211
Wailuku, HI 96793
(808) 249-8811 or (800) WHALE-11
outreach@pacificwhale.org
www.pacificwhale.org

REEF (Reef Environmental Education Foundation)

Armed with an underwater slate and pencil, and a good fish book, you can become an instant marine researcher. No degree, no former research expertise necessary—just a couple hours' training, and a passion for the sea.

Whether you're snorkeling or scuba diving, simply record the

Liz Foote

fish swimming around you when you dive Hawai'i's coral reefs. Keep track of how many individual species you see on the underwater paper REEF provides. After your dive, transfer your sightings to the REEF database online, or use the supplied Scantron form and send it to REEF headquarters in Florida. Your data will soon pop up on their Web site, along with the almost seven thousand REEF surveys that have been collected in Hawai'i since the program began in 2000.

And your findings are valuable. Scientists use the results of this underwater monitoring for marine conservation and resource management programs. The more surveys you do, the higher up you move on REEF's scale of expertise, meaning your dives take on even greater importance to the scientific community.

Best of all, you can do it on your own time, at any time—a small contribution to a large effort that couldn't exist without the help of so many hands and fins. To participate, membership in REEF is required and free. (Visit the REEF Web site to join—www.REEF.org.)

Maui's main REEF Field Station is Project S.E.A.-Link, which offers daytime field trainings or evening workshops in preparation for fish surveys. A related group is the community-based REEF club FIN, which also trains and surveys. On O'ahu, the Wild Dolphin Foundation acts as a local REEF Field Station, while also offering longer-term whale and dolphin research opportunities (see their listing in the O'ahu section). Visit the REEF Web site for a full listing of Field Stations across the state.

Project S.E.A.-Link

As its name implies, Project S.E.A.-Link makes connections—between scientists, teachers, nonprofit organizations, students, and the public. Their aim? To promote marine stewardship, while inspiring the next generation to care about and for the sea.

Liz Foote

Volunteer Activities: REEF surveys.

When: Field surveys at Honolua Bay and other locales year round. (Field trainings at least monthly, for 1/2 to 1 day. Evening trainings at least quarterly, for 2 hours.)

Who: Both individuals and groups welcome, all ages. Kids must be supervised by an adult. Participants must be experienced snorkelers or divers, and have their own equipment.

Hardiness Level (5 being most difficult): 1 to 2

Advance Notice: None needed.

Donations: Monetary donations are appreciated and tax deductible. Contact regarding equipment and supplies needs.

Contact: E-mail or phone.
Liz Foote
Project S.E.A.-Link
160 Kinohi Loa Loop
Wailuku, HI 96793
(808) 669-9062
Lfoote@projectsealink.org
www.projectsealink.org

FIN (Fish Identification Network)

The FIN folks describe themselves as fish nerds—they love REEF surveys so much, they do them together every month. You're welcome to join them, whether you're a snorkeler or diver. Membership is open and free.

Volunteer Activities: REEF surveys.

When: Once monthly (trainings and surveys on the same day), for 2 hours.

Who: Both individuals and groups welcome, all ages. Kids must be supervised by an adult. Participants must be experienced snorkelers or divers, and have their own equipment.

Hardiness Level (5 being most difficult): 1 to 2

Advance Notice: 1 day preferred.

Donations: Not accepted at this time. Ask about the Adopt-a-Reef program.

Contact: E-mail preferred.
Terri or Mike Fausnaugh
FIN (Fish Identification Network)
msfuzz@maui.net
www.projectsealink.org/forum/viewforum.php?f=27

"Participating in REEF surveys on Maui has been a wonderful way to deepen my understanding of marine life, and at the same time, give something back. I have been coming to Maui for over twenty-five years. But when I began to participate in fish counts shortly after I became a certified diver, I had an opportunity to become more than a visitor. I became part of the Maui community."

—Annette Lohman
Lakewood, California

Surfrider Foundation Maui Chapter

Surfrider Foundation—Maui

Surfrider volunteers on Maui's north shore extend their love of the ocean to include the shoreline surrounding Ho'okipa Beach Park. Renowned for its blustery trade winds and phenomenal waves, it attracts surfers, kitesurfers, and windsurfers from all over the world.

As with many places in the islands, native plants along this coastline are rapidly disappearing. Without them, the heavy rains create muddy runoff that deteriorates the nearshore waters and suffocates the reef with sediment. Naturally adapted to this dry coastal environment are native Hawaiian plants such as *hala, nanea, naupaka, pōhinahina*—they help secure the shore's sandy embankments, and slow the erosion that comes with gigantic north winter swells.

Surfrider Maui volunteers began stabilizing Ho'okipa's shoreline in 2002 by uprooting nonnative plants and replacing them with native species. It's an ongoing project that continually transforms this breathtaking beach park through the care of many helping hands.

(See Multi-Island section, under Surfrider Foundation, for more about the organization itself.)

Volunteer Activities: Clearing invasive plant species, installing and caring for native plants.

When: Year round (after school hours and on weekends), for up to 4 hours. (Contact for upcoming workdays; other days can be arranged.)

Who: Both individuals and groups welcome, all ages.

Hardiness Level (5 being most difficult): 2 to 3

Advance Notice: 1 week.

Donations: Monetary donations are appreciated and tax deductible. Contact regarding equipment and supplies needs.

Contact: E-mail or phone.
Jan Roberson
Surfrider Foundation, Maui Chapter
P.O. Box 790549
Pā'ia, HI 96779
(808) 298-8254
maui@surfrider.org
www.surfrider.org/maui

"It has been rewarding over the last several years to watch Ho'okipa be transformed into a park that is more pleasing to the eye. And it's rewarding to hear from residents that they see a positive change. The countless hours [spent] by hundreds of generous people are something the north shore should be proud of.

"Let's all hope the users of Ho'okipa continue to take ownership of this project in the years to come."

—Mark Rudd
Kula, Maui

Surfrider Foundation Maui Chapter

Ka Honua Momona

The tiny island of Moloka'i harbors more ancient coastal *loko i'a* (fishponds) than anywhere in Hawai'i. But the forces of nature and years of neglect have turned many into ruins, reducing once intricately laid rock walls to rubble.

In 2003, a group of Moloka'i residents formed Ka Honua Momona to address this troubling reality. They recognized that the many types of interrelated *loko i'a* were an ingenious way to sustainably manage an island's natural resources. Impressive feats of aquatic engineering, these ancient Hawaiian fishponds are made of massive rock walls that extend out into the sea and provided fish for Hawai'i's people. Ka Honua Momona still believes in this kind of sustainability.

Led by *kupuna lawai'a* (elder fisherman) Mervin Dudoit, the group manages two fishponds just east of Kaunakakai— Kaloko'eli Fishpond and Ali'i Fishpond. Mervin says he's out there most mornings if a willing volunteer wants to come lend a hand. Children are

> "The volunteering can be the physical labor of pulling invasive species, or it can be listening to *kūpuna* [elders] talk about their experiences fishing. It's a great place to get to know the culture, the fishponds, and each other.
>
> "My favorite part is pulling out the chain saw— it's a great way to get your aggressions out! And it's instant gratification, because now you can see where you've cleared.
>
> "Another favorite part is when we lay net, which we do to catch Australian mullet, an invasive species. It requires many hands. At the end of the workday is a potluck, and we fry up the mullet. It's a great way of connecting the whole idea—if you take care of the pond, the pond will feed you."
>
> —Todd Yamashita
> Kaunakakai, Moloka'i

invited too. "Kids playing in the pond moves the silt around," he points out. "That's good!"

Every third Saturday of the month, they also host Community Day at the ponds—working with volunteers to remove invasive mangroves, rebuild rock walls, and restore the ponds to a viable aquacultural resource. At the end of the workday, they often net the nonnative fish that have found their way into the ponds, and cook up a fresh fish feast. A tasty way to reinforce the idea that nature nourishes those who care for it.

Volunteer Activities: Fishpond work—removing invasive species and fixing rock walls.

When: Every third Saturday of the month; other days can be arranged.

Who: Both individuals and groups welcome. Kids must be supervised by an adult. (Camping may be available at the ponds—contact for details.)

Hardiness Level (5 being most difficult): 3

Advance Notice: 2 to 3 days preferred.

Donations: Monetary donations are appreciated and tax deductible.

Contact: Phone preferred.
Mervin Dudoit or Noelani Lee
Ka Honua Momona International
P.O. Box 482188
Kaunakakai, HI 96748
(808) 553-9968 or (808) 553-8353
khmi@mac.com
www.kahonuamomona.org

"Hawai'i's ancient fishponds are the cultural equivalent of Egypt's pyramids. Unfortunately, there is no state support for preserving them. Caring for these priceless cultural treasures has become the responsibility of nonprofit NGOs [nongovernmental organizations] like Ka Honua Momona, and the volunteers who support them. We do not want to see the last visible remnants of our ancient heritage disappear. . . .

"We run a small private high school, Ho'omana Hou, and our students conduct their science labs at Ali'i Fishpond. . . . Our students are fascinated by the pond's history and ecosystem. And standing on walls that were built hundreds or even thousands of years ago, they have learned to respect the engineering and aquaculture skills of their ancestors."

—Karen Holt
Moloka'i Community
Service Council
Kaunakakai, Moloka'i

Farm Apprenticeships

Developing an intimate relationship with the environment can take many forms. For some, it's a lifestyle—exposed to the daily rhythms of sunshine and clouds, storms and calm, volunteers learn hands on about all types of farming. In exchange, they earn room and board, which might mean living in a tent; often it means becoming part of a community. When many hands work together, it can be as nourishing as what the fields themselves provide.

There are many noteworthy farms in Hawai'i, yet too many to list in full here. So we're directing you to some reputable agencies that can help. How they work is by membership. Pay an annual fee and, in return, you'll get a comprehensive listing of participating farms—what they require, what you'll learn. After that, you're on your own to begin the discovery of farm apprenticing in the tropics.

WWOOF

WWOOF is like a password on the lips of agricultural types in the know. We've heard it as a noun (I'm part of WWOOF), an adjective (This is a WWOOF farm), a verb (I've WWOOFed before), and our favorite incarnation—the label (You're a WWOOFer? Me too!). The acronym means Willing Workers on Organic Farms, plus a few other translations you'll find on their Web site.

Relying purely on word of mouth for ten years running, the WWOOF Hawaii program attracts close to one thousand volunteers each year to the islands, some from as far away as Europe. Pretty impressive when you consider there are WWOOF organizations around the world. And a good reason to plan in advance for your dream Hawaiian destination.

Kirsten Whatley

The coordinators stress that experience is less important than attitude. So even if you've never volunteered on a farm before but would love to, here's your chance.

Volunteer Activities: Farm apprenticeships.

When: Year round, from several weeks to several months, depending on farm.

Who: Individuals or couples, ages 16 to 70. Younger kids may accompany adults if the host farm can accommodate them. No groups, due to accommodation difficulties.

Hardiness Level (5 being most difficult): 1 to 5

Advance Notice: 1 month.

Contact: Web site preferred. (Note: WWOOF Hawaii is coordinated from Canada.)
WWOOF Hawaii
4429 Carlson Road
Nelson, BC, Canada
VIL 6X3
wwoofcan@shaw.ca
www.wwoofhawaii.org

"It's taught me how to work—the environment demands it, and farming demands it, especially organic farming.

"In today's society, volunteering or interning redefines what it means to work for somebody. I think the concept of trade at this point in our capitalistic society could be a really interesting sociological experiment, because it's a different feeling trading for necessities like shelter and food than for pay. It's good."
—Jesse Clagett
WWOOFer,
Hāna Herbs and
Flowers
Hāna, Maui

HOFA

HOFA is Hawai'i's own Organic Farmers Association, and the state certifier of organic growers. They're a treasure-trove of information for those wanting to learn more about organic farming in the tropics. They even offer trainings on how to become an organic inspector.

Their network of participating farms gratefully accepts willing interns throughout the year. HOFA recommends doing a little research first—find out what crops will be in season at a particular farm, and you'll ensure a better match between your ideal learning situation and the apprenticeships available.

Craig Elevitch

Volunteer Activities: Farm apprenticeships.

When: Year round, from several weeks to several months, depending on farm.

Who: Individuals or families, age 18 and older.

Hardiness Level (5 being most difficult): 1 to 5

Advance Notice: Varies by farm.

Education: Annual membership meeting (usually in October) with guest speakers and informational tables is free and open to the public.

Contact: Web site preferred.
Maire Susan Sanford
Hawai'i Organic Farmers Association
P.O. Box 6863
Hilo, HI 96720
(808) 969-7789
hofa@hawaiiorganicfarmers.org
www.hawaiiorganicfarmers.org

OʻAHU

Candace Calloway-Whiting
Malama Na Honu

'Ahahui Mālama i ka Lōkahi

'Ahahui recognizes that Hawaiian culture is deeply linked to its native environment. You can see it in their logo—native plants, sea creatures, and wildlife encircle an ancient family petroglyph, symbolizing the relationships between plants and animals, land and sea, and the embrace of humans within them.

Embodied in this intricate motif are many layers of *kaona,* or hidden meaning. 'Ahahui invites you to come learn what they represent, while you help them preserve Hawai'i's native ecosystems and the living culture of its people. In their words: "As we see it, Hawaiian culture evolved in the embrace of native ecosystems, land and sea. As a result, Hawaiians developed a very intimate relationship with their natural setting . . . a deep love, respect, and knowledge of these places."

Their focus right now is Kawai Nui Marsh, Hawai'i's largest remaining emergent wetland and a primary habitat for native water birds. Sacred to Hawaiians, it is guarded by Nā Pōhaku o Hauwahine, an ancient rock formation of the Hawaiian *mo'o* (lizard) goddess.

At the edge of the marsh stands Ulupō Heiau, one of Hawai'i's first agricultural worship sites. Nearby, a system of terraces form ancient *lo'i kalo*

Rick Ka'imi Scudder

(taro patches), revealing a source of sustenance for one of the islands' largest ancient settlements.

Volunteering will mean helping to restore this ethnobotanical landscape, and in the process very likely restoring a part of yourself. ('Ahahui requests a small monetary donation from volunteers to help support their ongoing restoration work.)

Volunteer Activities: Restoration work—ethnobotanical, native forest, wetlands.

When: Usually every second and third Saturdays, from a few hours to 1 day.
(See www.ahahui.wordpress.com for upcoming projects.)

Who: Both individuals and groups welcome, age 10 and older.

Hardiness Level (5 being most difficult): 3 to 4

Advance Notice: Required—contact for details.

Education: Public tours of the Kailua *ahupua'a* (ancient land division) and Kawai Nui Marsh include topics of history, archaeology, native plant restoration, and birding. (Contact for upcoming dates, reservations, and donation fees.)

Donations: Monetary donations are appreciated and tax deductible. Contact regarding equipment and supplies needs.

Contact: E-mail or phone.
Kaimi Scudder
'Ahahui Mālama i ka Lōkahi
P.O. Box 751
Honolulu, HI 96808
(808) 593-0112
email@ahahui.net
www.ahahui.net

"It's very impressive to work at a place that is so ancient, and that figures in so many legends and chants, going back to about AD 300. It makes it much more than a plant restoration project. And that's really at the heart of 'Ahahui's work—the relationship to the land, not just in an economical or even an environmental sense, but in a spiritual sense.

"'Ahahui takes an environmental, educational, and cultural approach—all are equally important. It's very much a total experience."

—Waimea Williams
Kāne'ohe, O'ahu

Clean Water Honolulu

What we dump into O'ahu's storm drains doesn't go into the island's sewer system; it flows directly into the nearby streams and ultimately the sea. If polluted, it contaminates everything it touches—places where wildlife thrive, where people fish and play.

MAI KILOI 'OPALA
Dump No Waste

E Mālama I Ka Wai Ola
Protect our waters...
FOR LIFE

Iwalani Sato

There are over twenty thousand catch basins and inlets on the island, and more than 670 miles of storm drains. The task to keep them all clean is monumental. That's where volunteers become valuable assets. In just a few hours, a group can cover a slew of city blocks, stenciling storm drains with cautionary signs in both English and Hawaiian: *Dump No Waste. Protect Our Waters . . . for Life.* While they're there, they also collect trash and debris that would otherwise be destined for the ocean.

For volunteers who'd rather leave the city streets behind, there are stream cleanups around the island. (Local groups adopt a stream to caretake over a two-year period, and welcome participation on one of their scheduled cleanup days.) Recent projects took volunteers to Niu, Mānoa, Makiki, Kalihi, and Kapakahi streams, where they cleared the waters of debris so they could flow freely once again.

When waters are clear, the *'o'opu* (goby) return—an excellent biological indicator of a stream's health. Its image can be seen in the middle of the sidewalk stencil, a symbol of this ongoing effort to preserve our most fundamental resource.

Volunteer Activities: Storm drain stenciling, block cleanups, stream cleanups.

When: Year round, for 1 day per project. (Contact for scheduled dates.)

Who: Both individuals and groups welcome. For storm drain stenciling—age 9 and older. For cleanups—age 12 and older. (Kids under 18 must be supervised by an adult.)

Hardiness Level (5 being most difficult): 2 to 3

Advance Notice: 2 months.

Donations: Monetary donations are appreciated and tax deductible. Contact regarding equipment and supplies needs. Ask about the Adopt-a-Stream or Adopt-a-Block programs for your own group.

Contact: E-mail or Web site.
Iwalani Sato
City and County of Honolulu,
  Storm Water Quality Branch
Department of Environmental Services,
  Division of Environmental Quality
1000 Uluohia Street, Suite 303
Kapolei, HI 96707
(808) 768-3248
isato@honolulu.gov
www.cleanwaterhonolulu.com

"As a site leader and parent of a student at Washington Middle School, the [Makiki Stream] cleanup has evolved into more than what I expected. I do it for my kids. But, we have a really friendly group that has grown, including three generations of people who are working for the same thing, and new people who join us. . . .

"After five years, I feel a stronger sense of community; it's become a tradition."
—Laurence Kometani
McCully, O'ahu

Friends of Hanauma Bay

Hanauma Bay is incredibly popular. In the late 1980s, ten thousand people per day packed its sandy shores. That's a lot of trampling. And the bay was showing the wear—litter and marine hazards strewn every-where, fish being fed inappropriate  and harmful foods, damage to the precious coral and algae.

Candace Calloway-Whiting

In an effort to rescue this natural resource, the City adopted a far-reaching management plan that included putting a cap on daily visits to this nature preserve, dropping the numbers by two-thirds. It then shifted the bay's focus from recreation to marine education and preservation, requiring that all visitors watch a short film before beach access is granted. (In that handful of minutes, visitors learn how the bay was formed, how to interact with wildlife, and how to preserve the health and integrity of the reef.) The University of Hawai'i Sea Grant College was then contracted to develop and run its Hanauma Bay Education Program (HBEP), managing a cadre of volunteers and offering a wide range of award-winning learning opportunities for the public.

HBEP relies on its volunteer program, in which participants are asked to make no less than a six-month commitment. But shorter-term volunteering can be found through the Friends of Hanauma Bay, who run quarterly cleanups. Support their work and you will help further the collective mission of keeping Hanauma Bay wild.

Volunteer Activities: Nature preserve cleanups.

When: Quarterly, for 1 to 2 hours.

Who: Both individuals and groups welcome, all ages.

Hardiness Level (5 being most difficult): 1

Advance Notice: None needed.

Education: See Web site for upcoming events through the HBEP.

Donations: Monetary donations are appreciated and tax deductible. Contact regarding equipment and supplies needs.

Contact: E-mail harveys@hawaii.rr.com to receive announcements about upcoming cleanups.
Friends of Hanauma Bay
100 Hanauma Bay Road
Honolulu, HI 96825
www.friendsofhanaumabay.org

"The primary function is to pick up trash, but it's also to educate those who use the park and surrounding areas about the fragile environment and ecosystem.

"The benefit we gain as volunteers is it makes us aware of how we can contribute to improving and sustaining the environment in all its beauty. Without the cleanups we wouldn't appreciate having pristine shorelines as much.

"It gives you a sense of appreciation and accomplishment when you make a place better."
—Ryan Chang
Honolulu, O'ahu

Jeff Fukui

Hawaii Audubon Society

The Audubon Society has been around for over a century. In that time, they've been the leading voice for protecting wild birds and their native habitats across the nation.

In Hawai'i, this voice can't be loud enough. Of the ten most endangered birds in America, seven come from Hawai'i. The rarest of these is the

'alalā, or Hawaiian crow. Completely extinct in the wild, there are only about fifty in local captive breeding centers.

It would seem the birds are fighting a losing battle. Nonnative predators such as rats, cats, and mongeese prey on their nests, and introduced insects like mosquitoes spread avian diseases. The birds are also losing their habitat. Invasive plants are crowding out native species on every island, and introduced animals such as goats, pigs, and cows continue to degrade the birds' nesting areas.

The Hawaii Audubon Society is determined to reverse this distressing trend, and is fighting to restore native bird habitats throughout the islands.

With volunteer opportunities on O'ahu, their projects are ever changing. Contact them to find out which wild regions have the most pressing need for your help right now.

Volunteer Activities: Wildlife habitat restoration, administration, booth staffing.

When: Year round, for up to 3 hours per project.

Who: Both individuals and groups welcome, all ages.

Hardiness Level (5 being most difficult): 2 to 4

Advance Notice: Varies depending on event—contact for upcoming projects.

Education: Public field trips are offered monthly, educational lectures every other month. (Contact for details and donation fees.)

Donations: Monetary donations are appreciated and tax deductible. Contact regarding equipment and supplies needs.

Contact: E-mail or phone.
Hawaii Audubon Society
850 Richards Street, Suite 505
Honolulu, HI 96813-4709
(808) 528-1432
hiaudsoc@pixi.com
www.hawaiiaudubon.com

"A community of folks with a singular focus—to make wild bird habitats more accessible to birds and the humans who appreciate them . . . that's why volunteering is the right thing to do."
—Ann Egleston
Honolulu, O'ahu

"Representing the Hawaii Audubon Society at fairs, conferences, and other public events is always a great source of satisfaction to me. People respond so positively to the Society's beautiful images of the birds of Hawai'i, and we really enjoy sharing information about wild birds and their habitats."
—Wendy Johnson
Honolulu, O'ahu

Lyla Morgan

Hawai'i Nature Center—O'ahu

A safari into Makiki Valley to discover insects is the kind of fun the Hawai'i Nature Center has on weekends or whenever school is out. As is making art from tree rubbings. And racetracks for slugs. And masks decorated with spider designs for Halloween. Some days, kids and their families are guided through the Pouhala Marsh watershed, where endangered wetlands birds still soar and sing.

It takes a lot to keep the forest from reclaiming the cleared spaces of nature's classroom. Help with vine eradication, landscaping and gardening, or restoration of the *lo'i kalo* (taro patch) is always welcome. Or join their monthly Pouhala Marsh workdays, where volunteers maintain habitat areas for wetlands birds, particularly the endangered *ae'o* (Hawaiian stilt).

(See Multi-Island section, under Hawai'i Nature Center, for more about the organization itself.)

Volunteer Activities: Wetlands restoration, vine eradication, *lo'i kalo* (taro patch) restoration, landscaping, gardening.

When: Most activities—year round. Pouhala Marsh workdays—one Saturday per month, for 2 hours.

Who: Both individuals and groups welcome; groups preferred. Age 7 and older; kids under 16 must be supervised by an adult.

Hardiness Level (5 being most difficult): 3 to 4

Advance Notice: Pouhala Marsh workdays—3 days. Other activities—2 weeks.

"Pouhala Marsh is located between Waipahu town and Pearl Harbor. Unfortunately, this area is populated and suffocated by unnecessary plants such as cattail and mangroves. . . . Similar to mangroves, cattails prevent the healthy growth of the wetlands. This is where the Hawai'i Nature Center comes in. They preserve the wetlands, for they are important to the living creatures that decide to inhabit these areas. . . .

"I find that it is really important to take care of our land, and these projects with the Hawai'i Nature Center allow me to be really grateful for what our ancestors have done for our land. Participating in this project makes me feel accomplished and good about myself. I feel like I've done something good for this land, and I've learned that every helping hand counts."

—Krystina Begonia
Student,
Maryknoll School
Kapolei, O'ahu

Education: Weekend family programs and hikes, and extensive kids' programs are available. (See Web site for details and fees.)

Donations: Monetary donations are appreciated and tax deductible. Contact regarding equipment and supplies needs.

Contact: E-mail preferred (put Hawai'i Nature Center Volunteer Request in subject heading).
Pauline Kawamata
Volunteer Program Manager
Hawai'i Nature Center
2131 Makiki Heights Drive
Honolulu, HI 96822
(808) 955-0100 or (888) 955-0104
volunteer@hawaiinaturecenter.org
www.hawaiinaturecenter.org

"I don't remember why I first volunteered, but every day I'm glad I did. I get kids out on field trips who have no idea there's a forest and wilderness on O'ahu. They wander around with their eyes wide open. They're astonished.

"I'll have the kids do activities that seem like games to them, but they're actually learning science, whether ecology or geology or biology. For example, I'll have the kindergartners make mud pies: What's the first ingredient? I ask. Soil. So we sprinkle in crumbs of dirt . . . then dead leaves . . . the last ingredient is rain, so we sprinkle in the water together. Then they reach out their little hands, and I give them each a ball of the mud we've just made. They pat it out and put some sprinkles of seeds on it. Now they have soil cookies.

"I say, Are we going to eat this? Ugh! No! Then I ask, So who does eat dirt? And each child picks out a special plant and feeds their cookie to it. They've just had a lesson in soil ecology, but to them it's playing.

"Once you find something you really enjoy, it's wonderful to be able to contribute to your community in that meaningful way."
—Jeanne Davis
Wai'alae Iki, O'ahu

Honolulu Zoo Society

They started as the Zoo Hui in 1969, set up to support the denizens of the Honolulu Zoo and the work of those who care for them. Today, they're called the Honolulu Zoo Society, with a membership of nearly eight thousand animal lovers throughout Hawai'i and the continental United States.

The Society helps educate the public about wildlife and conservation, and helps zoo staff ensure the places these wild animals call home stay in prime condition. And they couldn't do it without their volunteers.

Each year, these selfless souls donate over twenty thousand hours to help keep the zoo running smoothly. Some spend months at a time caring for the animal residents, such as Kruger, the white rhinoceros, or Cobb, the corn snake. Other volunteers come in groups for day projects, like making children's masks that look like animal faces for the Society's special events, or helping maintain the zoo's habitats, or beautifying the grounds.

Photo by Dollye Drum

Still others become ZooParents, adopting a bird, reptile, or mammal, including threatened or endangered species—the animals stay at the zoo, the adoptive parents love them from afar. Or in the Society's words: "You provide the love, we provide the expert care."

With an emphasis on tropical ecosystems, the Honolulu Zoo and its supporters provide one more way to learn about and care for wildlife in Hawai'i.

Volunteer Activities: Animal keeper assistant, assistant in Keiki Zoo, beautification, maintenance, special events.

When: Zoo assistants—year round, 3-month minimum commitment, for 1 day per week. Beautification/maintenance/special events—year round, for 1 day per project.

Who: Zoo assistants—individuals only, age 18 and older for animal keeper assistants; age 16 and older for assistants in Keiki Zoo. Beautification/maintenance/special events— groups only (of at least 5 people); kids must be supervised by an adult.

Hardiness Level (5 being most difficult): 1 to 4

Advance Notice: Zoo assistants—interviews by appointment or on Walk-in Wednesday mornings, when volunteer coordinator accepts walk-in interviews (free zoo admission given to interviewees). Beautification/maintenance/special events— 1 month.

Education: The Society offers a wide range of programs for families and school groups,

"If you told me two years ago that I would be working with llamas, goats, pigs, chickens, lizards, and all other sorts of animals, I would not have believed you. . . .

"As a volunteer in the Keiki Zoo, I have so much fun doing the so-called dirty work; I mean how could you not have fun washing Lani Moo, the cow? . . . Working with Cobb [the corn snake] is really special to me, because not many people on the island can say they've seen or even touched a snake, and I have assisted [with] countless snake encounters.

"I know I am there to help the animals and the zoo staff out, but honestly I feel like I get more out of it, for it teaches self-discipline, communication skills, commitment, work ethics, responsibility, and so much more. It takes me six buses and four hours to get to the zoo and back, but it is worth it. I would not trade this experience for anything!"

—Lauren Chavez
Student,
Kapolei High School
Kapolei, O'ahu

including Breakfast with a Keeper, Twilight Tours, and Snooze in the Zoo (after-dark tours that culminate in roasting marshmallows around a campfire). (See Web site for details and fees.)

Donations: Monetary donations are appreciated and tax deductible. Ask about the Animal Adoption program.

Contact: E-mail preferred.
Barbara Thacker
Honolulu Zoo Society
151 Kapahulu Avenue
Honolulu, HI 96815
(808) 926-3191, extension 11
bthacker@honzoosoc.org
www.honzoosoc.org

"We have been volunteering at the Honolulu Zoo for a decade, and without any reservations it has been a wonderful and enriching experience for us. . . . Every animal has a different personality and no two volunteer days are the same. We get dirty, we get tired, and we enjoy every minute of it.

"Even after all these years, we never fail to look at each other while driving home and say, 'Isn't it just amazing that we get to do what we do?'"

—Dr. Dave and
Bridgit Stegenga
Kailua, O'ahu

Photo by Barbara Thacker

Ho'oulu 'Āina
Kalihi Valley Nature Park

'O ka hā o ka 'āina ke ola o ka po'e. "The breath of the land is the life of the people." At Kōkua Kalihi Valley, they take this seriously. They believe the land and our bodies are one and the same—when we restore health to the land, we heal our own bodies. Ho'oulu 'Āina is one special place where this connection is happening.

On ninety-nine acres in the back of Kalihi Valley, members of the nonprofit health organization Kōkua Kalihi Valley are reforesting the uplands with native species; they're restoring ancient agricultural terraces and developing community gardens. Their vision is to provide a gateway for the community to live healthier, more active lives—growing their own food, reforesting the landscape, trailblazing and then using the paths for hiking and walking.

Volunteers are invited to become part of this transformation. Help is needed with gardening, forestry, trail building, and construction. You can pitch in at their monthly community workdays, or arrange a time to come during the week.

As coordinator Puni Kukahiko describes it, "You can expect to grow and have fun as we share in sweat and laughter and probably a little rain."

Volunteer Activities: Gardening, forestry, trail building, construction.

When: Community workdays—every third Saturday morning (see www.hoouluaina.blogspot.com for upcoming dates). Monday through Friday mornings also available (contact to schedule).

"It was truly an amazing experience. I had no idea the kind of work Puni and the others do out there, but was really impressed by it. . . . I believe the labor and efforts they do for this community are imperative and sometimes overlooked in today's society.

"Personally it was great for me to work side by side with the teens . . . they definitely took pride in the work they did, and I could sense a feeling of accomplishment among the group. I believe the teens and youth in this community need to do more activities like this, which allow them to take responsibility and ownership of the land. . . . We definitely achieved it that day."

—Erin Berhman
Kailua, O'ahu

"The environment is peaceful and inviting. Most of the time I forgot I was working because we were having so much fun. . . .

"The biggest thing we learned was that it takes a lot of manpower, volunteerism, and dedication just to maintain a balance in nature."

—David Lau
Waimānalo, O'ahu

"I wanted to find a way to give back to my community and the land (island). . . . We put on gloves, hiked up the hillside, and then pulled and shoveled out invading plants. . . . Then we planted sweet potatoes, which help to keep out the invasive species. It rained plenty, but we had fun working in the mud and laughed a lot. . . .

"[The most rewarding part was] the feeling of accomplishment and the experience of working 'hands on' [with] the land. I worked so hard that day I lost a couple of pounds."

—Delio Agbayani
Honolulu, O'ahu

Who: Both individuals and groups welcome. Kids under 14 must be supervised by an adult.

Hardiness Level (5 being most difficult): 2 to 5

Advance Notice: Individuals—1 week; groups—1 month.

Donations: Monetary donations are appreciated and tax deductible if made to Kōkua Kalihi Valley. Ask about the Adopt-a-*Koa* program.

Contact: E-mail preferred.
Puni Kukahiko
Ho'oulu 'Āina
Kalihi Valley Nature Park
3659 Kalihi Street
Honolulu, HI 96819
(808) 841-7504
puni@kkv.net
http://search.volunteerhawaii.org/org/16882068.html

Lyon Arboretum

A university garden in a rain forest? Only in Hawai'i.

On nearly two hundred acres at the upper end of Mānoa Road, the Lyon Arboretum has presided over the landscape since 1918. In those early years, it was a bit smaller and served as a forest restoration project, an attempt at reviving a terrain that had been devastated by cattle.

The trees came first, about two thousand of them. Later two thousand or so plants were introduced. Today it's a thriving center for the preservation and cultivation of rare and endangered native Hawaiian species.

By definition, an arboretum is a place where trees and plants are cultivated for science and education. In essence, it's a living library—for scientists, students, visitors, community, and volunteers.

Lyon Arboretum

Opportunities run the gamut, from helping with the ethnobotanical garden, exotic tropicals, native plants, herbs, or the children's garden, to removing invasive weeds, to greeting visitors at the garden gate. Or take your morning walk through the arboretum and become a "trail sweeper," clearing debris from your path as you go.

There are also sporadic opportunities such as harvesting seasonal fruit for the Hawaii Foodbank, working with photo archives, or helping with special events like plant sales.

Volunteer Activities: Plant/garden/grounds maintenance, invasive weed removal, garden greeter, trail sweeper, special projects.

When: Most activities—1 day (more preferred). Garden greeter—1-month minimum commitment. (Contact for current season's special projects.)

Who: Both individuals and groups welcome. Kids under 18 must be supervised by an adult.

Hardiness Level (5 being most difficult): 1 to 5

Advance Notice: Required—varies depending on project.

Education: Docent-led garden tours, adult and children's classes, school field trips, and internships are offered. (See Web site or contact for details and fees.)

Donations: Monetary donations are appreciated and tax deductible. Contact regarding equipment and supplies needs. (Direct all donations questions to Dr. Christopher Dunn.) Ask about the Adopt-an-(Endangered)-Plant program.

Contact: E-mail preferred.
Jill Laughlin
Lyon Arboretum
University of Hawai'i at Mānoa
3860 Mānoa Road
Honolulu, HI 96822
For volunteering or educational programs—
(808) 988-0461. For donations—(808) 988-0456.
vollyon@hawaii.edu
www.hawaii.edu/lyonarboretum

Lyon Arboretum

Candace Calloway-Whiting

Mālama Na Honu

When you come to the North Shore, you might meet Brutus, Honeygirl, or Olivia Dawn. They're *honu*, turtles, Hawaiian green sea turtles to be exact. And one of them (or their cousins) is usually basking on the sands at Laniākea Beach.

Unfortunately, some who come to visit them are less than respectful. Beachgoers have been known to drag the turtles from the ocean, sit on them, lift or manipulate them for a better picture, or try to turn them over.

In 2005, the head of turtle research at the NOAA (National Oceanic and Atmospheric Administration) Pacific Islands Fisheries Science Center decided enough was enough, and launched the Show Turtles Aloha campaign. Volunteers stepped forward and began educating viewers on how to respect these threatened (meaning "likely to become endangered") creatures. This laudable effort has since been turned over to the newly formed Mālama Na

"I feel like as an advanced species we have a responsibility to help right what we have wronged. I want to make a difference, at least if only a little, and I'd like to educate people. I feel like that is one of the key factors in turning around the decline of a species. . . .

"I think the biggest thing I have learned is not to make assumptions about the people that come to see the turtles, and to not make assumptions about the turtles either! They are great survivors, they are wild animals, and the interface between us and them in their environment is always a good reminder that we are still animals too, as much as we'd like to think we're not."

—Mariah Snipes
Wahiawā, O'ahu

"It is difficult to describe what it is like volunteering at Laniākea. I have met many great passionate people/volunteers, and my time at Laniākea is so relaxing and enjoyable. Volunteering has been a stress reliever for me.

"The *honu* are fascinating animals, and I enjoy talking with the different beachgoers that have ventured to Laniākea to see [them]. . . . It never gets old watching the *honu*."

—Sarah Svedberg
Hale'iwa, O'ahu

Honu Foundation, who trains volunteers for day shifts of public outreach and turtle protection.

While the turtles nest and breed five hundred miles northwest at the remote French Frigate Shoals, they've chosen Laniākea as a favorite resting and feeding spot. It's our responsibility to share it with them respectfully.

Volunteer Activities: Green sea turtle protection and public education.

When: Year round, for 2- to 3-hour daytime shifts, 2-month minimum commitment. (Free volunteer training includes 1 hour of instruction and 2 sessions shadowing a more experienced volunteer.)

H. Patrick Doyle

Who: Individuals preferred, though groups welcome. Kids under 18 must be supervised by an adult.

Hardiness Level (5 being most difficult): 3

Advance Notice: 3 to 4 weeks.

Donations: Monetary donations are appreciated (contact about tax-deductible status).

Contact: E-mail preferred.
Mālama Na Honu Foundation
P.O. Box 1078
Hale'iwa, HI 96712
info@malamanahonu.org
www.malamanahonu.org

Mālama Na Honu

"When I learned of the Honu Guardian Program, I thought this was the ideal way for me to give back something, not only to the wonderful creatures, but to the islands, its waters, and its people that have provided me with such a wonderful life and so much enjoyment. I had no idea that I would become as passionate as I have about our *honu 'ohana* [turtle family] at Laniākea. And how important it has become for me to ensure that they are well-protected, and be a voice for them to ensure that protection. . . .

"The most rewarding part of my volunteer experience is by far the encounters I have with children on the beach. . . . After all, they are the next generation to inherit all of this. If we can instill in them an importance of respecting, protecting, and conserving our planet, perhaps in the future we will see that we can correct some of the damage we have done in the past, and at times are still doing in the present."
—Patrick Doyle
Waikīkī, O'ahu

Photo by Mashuri Waite

Mānoa Cliff Trail Project

Mere minutes from the bustle of Honolulu, on the cliffs of Mount Tantalus's east side, volunteers are quietly saving a forest. They're saving the native *koa*, *ʻōhiʻa*, and *māmaki* trees and a rainbow of indigenous plants. They're preserving the habitat for native birds such as the *ʻamakihi* and *ʻapapane*, and unusual native insects like Hawaiʻi's crickets (the world's fastest-evolving invertebrates) and carnivorous, insect-eating caterpillars.

In an area already rich in native species, most trail work involves pulling and cutting alien weeds to give more room to the original dwellers, to let the light back in and allow the rain to nourish the plants' tender leaves and roots.

Volunteers work as a team, with Mānoa Valley just over their shoulders, its lacy waterfalls drawing the eye to the valley's forested back wall. Out toward the horizon, the sweeping vista encompasses Honolulu down by the sea and the drama of Diamond Head beyond. By the end of the morning, the workers can leave satisfied knowing that a little more of the indigenous forest has been preserved.

In the words of volunteer Kari Benes, "During the hikes the leaders share an earnest appreciation for and knowledge of the native plant species that spur energy in any volunteer to weed out the invasive species. Volunteering . . . is very rewarding, because after only a couple hours you can see the difference one group can make."

Volunteer Activities: Native forest restoration.

When: Twice monthly, usually the first and third Sunday mornings, for 4 to 5 hours.

Who: Both individuals and groups welcome, all ages. Kids must be supervised by an adult.

Hardiness Level (5 being most difficult): 3

Advance Notice: Preferred for groups of 6 or more.

Donations: Not accepted at this time.

Contact: E-mail preferred.
Mānoa Cliff Trail Project
c/o Conservation Council for Hawai'i
P.O. Box 2923
Honolulu, HI 96802
(808) 593-0255
manoacliffnatives@gmail.com
www.conservehi.org

"Helping in the Mānoa Cliff Trail Project is for me an excellent way to contribute to the conservation of native plants in the place I am living. . . . Besides having a good view of Mānoa and Honolulu, getting out of the city environment, talking to nice people, and feeling useful in [a] practical conservation effort, it is a great way of learning about Hawaiian native plants and the invasive species found around here. The group has impressive, knowledgeable people!"
—Isabel Belloni Schmidt
PhD Student,
University of Hawai'i
at Mānoa
Honolulu, O'ahu

"This is a real do-something project. We get our hands dirty and sweat a bit to help out the forest, and we see the fruits of our labor.

"Everyone has talents that they lend to the project, and we learn together as we go along. We talk story, eat, and enjoy each other's company."
—Glenn Metzler
Honolulu, O'ahu

"Why do we volunteer with Makiki WAI? It's not because it's particularly easy or fun. It's because it is the right thing to do. By virtue of our presence here, we and our families are part of this *'āina* [land], and it is a part of us.

"This land and its unique plants have evolved over millions of years, and in the short time since we humans have arrived, we've managed to nearly destroy it. That which we haven't destroyed, we are endangering. So volunteering is one way to care for, nurture, and begin to restore our *'āina*.

"For me, it is also a matter of faith and belief that what we do today—good and bad—will have an effect in the future. The native plants we carefully nurture will reproduce and reclaim their rightful and natural habitat. They certainly won't if we do nothing."

—Tom Rau
Kāne'ohe, O'ahu

Na Ala Hele

Na Ala Hele—O'ahu

The *wai* of Makiki WAI stands for "Watershed Awareness Initiative." But *wai* also means "water"— an apt acronym for Na Ala Hele's restoration work in the Makiki Valley.

A watershed is the land that contains a set of streams that feed into a larger waterway, such as a bigger stream, a lake, or the ocean. The aim of Makiki WAI is to restore a segment of the area's watershed, then use that segment to educate others about the importance of watersheds, teaching how they sustain life and contain our most valued resource.

By association, the Hawaiian concept of *ahupua'a* surfaces. This traditional way of dividing up the islands provided people with the wealth of both land and sea—starting in the mountains, it stretched down in a pie wedge shape to and including the ocean. While *ahupua'a* is a cultural term and *watershed* a scientific term, they frequently

follow the same lines on the land and are often used interchangeably.

Volunteers at Makiki are helping clear invasive species from the forest and replanting natives, improving portions of the trail and building bridges, and installing educational displays and signs. By returning the region to its natural state and stabilizing the soil with native plantings, you'll help to reduce erosion and improve water retention, restoring health to the entire ecosystem.

(See Multi-Island section, under Na Ala Hele, for more about the agency itself.)

Volunteer Activities: Replacing invasive species with natives, improving trails, installing displays and signs.

When: Monthly, usually the fourth Saturday, for 5 hours.

Who: Both individuals and groups welcome. Kids under 13 must be supervised by an adult.

Hardiness Level (5 being most difficult): 1 to 3

Advance Notice: None needed for individuals; groups—1 week.

Donations: Monetary donations are appreciated and tax deductible. Contact regarding equipment and supplies needs.

Contact: E-mail preferred.
Aaron Lowe
Department of Land and Natural Resources
Na Ala Hele—Hawai'i Trail and Access System
2135 Makiki Heights Drive
Honolulu, HI 96822
alowe@hawaii.rr.com
www.hawaiitrails.org

"For two years I was a regular volunteer for the Makiki Wai watershed protection project, while keeping an extremely busy schedule as a graduate student at the University of Hawai'i at Mānoa.

"Every month this activity refreshed me and gave me a sense of grounding (literally!) with this land and community. I relished the opportunity to escape from my studies for a half day to maintain trails and cultivate a garden of native plants, and I felt great satisfaction when I saw how the fruits of our efforts transformed the land into an educational and recreational resource for the community.

"By volunteering our creative energies, we can minimize the ills of urban living, and create healthy links between our forests, our waters, our communities, and our own lives."

—Robban Toleno
Marlboro, Vermont

Nani 'O Wai'anae

"Our main job here at Nani 'O Wai'anae is to educate our neighbors on how and when to reuse, reclaim, and recycle. We go out practically every weekend to reclaim tires from the bushes and gather bottles, cans, and rubbish from the roadways, and always we encounter people who are curious and want to know more.

"[On] one project, we reclaimed over eight hundred tires and hundreds of bags of litter, and involved over fifty people from the community. That day we knew that we had made a real dent in the problem and showed the residents in the area how to go about it."

—Judi Duffy
Wai'anae, O'ahu

The O'ahu affiliate of the nationwide Keep America Beautiful stays busy by rounding up old tires, painting park tables, cleaning up graffiti, and picking up trash. As volunteer Lori Gossard puts it, "It seems crazy to have so much fun picking up other people's litter. It's like eating peanuts: the more you pick, the more you want to."

With projects happening two to three times a month, you can have a lot of "crazy fun" helping keep O'ahu beautiful.

(See Multi-Island section, under Keep America Beautiful, for more about the national organization.)

Volunteer Activities: Cleanups and beautification.

When: 2 to 3 times per month, usually Saturdays.

Who: Both individuals and groups welcome. Kids must be supervised by an adult.

Hardiness Level (5 being most difficult): 1 to 5

Advance Notice: 1 to 2 days.

Donations: Monetary donations are appreciated and tax deductible. Contact regarding equipment and supplies needs.

Contact: Phone preferred.
Katy Kok
Nani 'O Wai'anae
84-183 Makau Street
Wai'anae, HI 96792
(808) 696-1920
naniowai@lava.net
www.kab.org

Nani 'O Wai'anae

O'ahu Invasive Species Committee

Some call a weed just a plant growing in the wrong place. But when that plant wears out its welcome and finds its new surroundings just like home—or better than home—it starts to get greedy.

It steals sunlight from the host culture of plants, and inhibits their right to rain. It

O'ahu Invasive Species Committee

draws more and more nutrients from the soil to feed its aggressive growth. The new kid on the block, in essence, becomes the bully.

Invasive Species Committees are set up on all main Hawaiian Islands, with regular volunteer opportunities on O'ahu. Their goal is straightforward—eliminate the invaders before they take over.

For most projects, you'll likely have to hike a small distance to reach the designated wilderness site. Once there, you might find yourself pulling Himalayan blackberry from the native forest, or eradicating the *manuka* plant from a *koa* tree habitat, or stopping the spread of fountain grass down at the coast.

In partnership with organizations around the island, the O'ahu Invasive Species Committee (OISC) often works at sites that are not just ecological but cultural in significance. Your volunteering helps to perpetuate and strengthen this link between Hawai'i's environment and its native heritage.

Volunteer Activities: Invasive species removal. *(This may involve applying over-the-counter herbicides. If you prefer not to use herbicide, let the staff know—they have many other tasks available.)*

When: Every second Saturday, for 1 day. (Other days can be arranged—contact to schedule.)

"I have been after miconia [a highly invasive tree] for almost twelve years, and I would like to see it gone for good. Hawai'i is looking like the average South American country with all their trees, shrubs, heliconia, and grasses invading our forests and mountains. We are desperately trying to rescue our native plants and trees from extinction. To do that we need to [protect] their habitats from invasives such as miconia or they will never survive."
—Charlotte Yamane
Kāne'ohe, O'ahu

"I enjoy volunteering with OISC because more than just being out enjoying the environment, I get to help restore Hawai'i's ecosystem by removing invasive species.

"I keep coming back because I'm crazy about the Hawaiian environment, enjoy getting dirty, and love fluorescent colors." [Volunteers wear bright shirts to help see each other in the forest.]
—Patrick Chee
East Honolulu, O'ahu

Who: Both individuals and groups (by special arrangement) welcome, age 14 and older.

Hardiness Level (5 being most difficult): 5

Advance Notice: 1 day.

Education: Information about species invasive to Hawai'i is available on their Web site.

Donations: Not accepted at this time.

Contact: E-mail or phone.
Julia Parish
O'ahu Invasive Species Committee
2551 Waimano Home Road, Building 202
Pearl City, HI 96782
(808) 286-4616
oisc@hawaii.edu
www.oahuisc.org

O'ahu Invasive Species Committee

Sierra Club—O'ahu

The Sierra Club's O'ahu group runs service projects ranging from one day to one week (one-day projects on O'ahu are free; multiday projects have fees to cover airfare to neighbor islands, meals, lodging, etc.). The off-island trips might land you on Moloka'i, where volunteers can help with *nēnē* (endangered Hawaiian goose) habitat recovery, alien species removal in a mountain forest preserve, or on the remote Kalaupapa Peninsula. The O'ahu group also invites assistance with mailings, creation of educational displays, and other specialty projects. Let them know your talents.

(See Multi-Island section, under Sierra Club—Hawai'i Chapter, for more about the organization itself.)

"Most of the volunteer work I do with the Sierra Club involves the cleanups. . . . On occasion I lead hikes, but I find the cleanups very rewarding. We live in such a beautiful place, it's a shame to see garbage lying around spoiling the beauty. I also like to get the garbage before it gets into the ocean and affects the seabirds and sea life.

"I also like the three-day service projects. They're a lot of fun. You get to visit and stay at places the general public isn't allowed to, and be outdoors planting or clearing the land.

"I like to think that the work I volunteer for will make a difference, and that any improvements to the land will be around so that future generations can appreciate it."
　　　　　—Deborah Blair
　　　　　Honolulu, O'ahu

"I was raised with the idea that you have to give back to your community. It was natural then that if you wanted to hike, you had to help protect the places you wanted to do it in. It's very satisfying. Over the years, you see it does make a difference. I see it as a way for people to work together and cooperate, to accomplish something—I think the world needs a lot of this.

"I think that women, in particular, and men too, need to get dirt under their fingernails. I think there's a chemical reaction that goes on with soil, your body, and sunshine that just feels good. We're always walking around on sidewalks and elevators, and we're not in touch with nature.

"All of my friends now are friends that I made through the environment somehow—someone I met on a service project, someone I met through the Sierra Club. These are friends that you have for a long, long time. And there are lots of things you can do together, because you have something in common."

—Annette Kaohelaulii
Kāne'ohe, O'ahu

Volunteer Activities: Removing invasive species, planting natives, trail clearing and maintenance, cleanups, specialty projects.

When: Year round, project lengths vary.

Who: Both individuals and groups welcome, all ages.

Hardiness Level (5 being most difficult): 1 to 5

Advance Notice: See Web site.

Education: Outings such as hikes above Hanauma Bay to view rarely seen native plants, into the shady depths of Mānoa Valley, or up to the Ko'olau summit with Windward O'ahu falling away dramatically beneath your feet. They also host free presentations and monthly "beer nights" to inspire discussions on timely environmental topics. (See Web site for details.)

Donations: Monetary donations are appreciated and tax deductible if made to the Sierra Club Foundation (contact office for more information). Contact regarding equipment and supplies needs.

Contact: E-mail preferred. For service projects—Randy Ching, oahurandy@yahoo.com. For other volunteer opportunities—Elizabeth Dunne, scvolunteers@gmail.com. For outings—Gwen Sinclair, gsinclai@hawaii.edu.
Sierra Club—O'ahu Group
1040 Richards Street, Room 306
Honolulu, HI 96813
(808) 538-6616
www.hi.sierraclub.org (click on O'ahu Group)

Surfrider Foundation—Oʻahu

Rising to meet challenges, choosing actions with intention, learning to fall—this is some of the wisdom surfing teaches. If you can master (or even brush up against) these lessons in the water, it can change your life.

This is the theory behind Spirit Sessions, surfing lessons for at-risk girls led by Surfrider Oʻahu volunteers. Inspired

Courtesy of the Surfrider Spirit Sessions. Photograph by Marvin Heskett.

by program leader Cynthia Derosier's book *The Surfer Spirit*, surfers become mentors, teaching the fundamentals of finding one's strength and focus in the ocean, acceptance and calm. As volunteer Annabel Murray puts it, "The ocean is a place of refuge, of peace, of power, of ever-shifting beauty. Each girl is learning how to utilize the force of the ocean in her life."

The Oʻahu Surfrider's schedule also makes room for monthly beach cleanups around the island. And they welcome help with their administrative and fund-raising tasks. Any and all can join these ongoing efforts to support Oʻahu's shores and the oceangoers who use them.

(See Multi-Island section, under Surfrider Foundation, for more about the organization itself.)

Volunteer Activities: Surf lesson mentorships, beach cleanups, administration/fund-raising.

When: Surf lesson mentorships—Saturday mornings, for 4 hours, 10-week minimum commitment. Beach cleanups—usually last Saturday of each month, for 1 to 2 hours. Administration/fund-raising—ongoing.

"I've been a Spirit Session mentor from the beginning of the program. Cynthia and her book, *The Surfer Spirit*, have provided a wonderful foundation for surfing and spending time with the amazing girls of 'Girl's Court.' . . . Some girls have become aware of untapped strength, surfing waves they never imagined themselves riding with such freedom. Some girls have acknowledged their fears, and are slowly allowing themselves to test the waters and themselves; a wet foot an achievement, a briefly submerged face a triumph.

"It is wonderful to see early morning grogginess and resistance turn to glowing faces and excited renditions of waves missed and waves caught, 'surf stories' that lay a continuing foundation for a life of pushing physical and mental boundaries.

"The Spirit Sessions, for me, have been about watching these young women connecting with the ocean, with each other, and with themselves."

—Annabel Murray
Honolulu, O'ahu

Who: Surf lesson mentorships—female surfers with intermediate surf skills (able to catch 4-foot faces and turn); other water sports and safety skills a plus; must be comfortable teaching others; serving as a mentor outside of sessions also encouraged. Beach cleanups—both individuals and groups welcome, all ages. Administration—individuals with basic office and/or fund-raising skills.

Hardiness Level (5 being most difficult): Surf lesson mentorships and beach cleanups—2 to 3. Administration—1.

Advance Notice: Surf lesson mentorships—3 to 4 weeks before next session (contact for schedule). Beach cleanups—none needed. Administration/fund-raising—contact for details.

Donations: Monetary donations are appreciated and tax deductible. Contact regarding equipment and supplies needs.

Contact: For surf lesson mentorships—Cynthia Derosier, www.thesurferspirit.com/spirit.htm. For beach cleanups—Doug Rodman, (808) 637-4151. For administration/fund-raising or general questions:
Surfrider Foundation, O'ahu Chapter
c/o Scott Werny
1917 Kuapapa Place
Honolulu, HI 96819
Scott Werny, clearwater@hawaii.rr.com
Marvin Heskett, mhesketts@mac.com,
   (808) 728-4617
www.surfrider.org/oahu

Wild Dolphin Foundation

Spinner dolphins are morning creatures. That's when the whole pod comes together to play and twirl above the water like airborne tops. This is also the time when Hawai'i's dolphins are spotted caressing each other with their tail flukes, or swimming belly to belly, one under the other; some even appear to hold hands.

As the sun rises higher in the sky, the dolphins draw closer for a collective nap. Yet researchers have learned that only part of their brain rests at any given time. With their sonar switched off, they rely instead now on their eyes—a whole school of eyes working together—to detect predators, particularly hungry sharks. In the afternoon, they leave the shelter of the bay to enter deeper waters to fish and feed. And the cycle starts again.

In the living laboratory of Hawai'i's ocean, researchers are able to opportunely study these behaviors every day. That's the approach of the Wild Dolphin Foundation (WDF)—to understand these sea mammals in their own environment, to educate others about why they're so special, and to preserve

Wild Dolphin Foundation

> "What's it like working out there? Wonderful. Every time I went out, I found the staff to be friendly, knowledgeable, and capable. The Wild Dolphin Foundation has everything so well-organized and thought-out that it is easy to concentrate on the job at hand. . . .
>
> "I've learned that it's possible to combine vacation and volunteerism in a truly meaningful way. And I've met people who are genuinely dedicated to both the animals and to enhancing our knowledge and understanding of them.
>
> "Through volunteering with Wild Dolphin, I also discovered the value of conscientious ecotourism, of visiting animals in their own habitats, and the gifts we receive by learning how to enter their world and encounter them on their terms."
>
> —Candace Calloway Whiting
> Seattle, Washington

their habitat for future generations. (Their work also includes whale study.)

Volunteers who can commit at least two months' time are invited to join a network of field researchers. The ideal candidate has either a background in or a passion for wildlife, conservation, (marine) biology, and/or environmental science. You must be willing to work both with a team and alone, as you might find yourself at times the sole researcher on a boat.

If you'd rather contribute in the short term, WDF is also a REEF Field Station. After a brief training, you'll collect data on the underwater ecosystem for use by marine researchers. (*See Multi-Island section, under REEF, for details on how these fish surveys work.*)

Volunteer Activities: Boat-based dolphin and whale research; REEF fish surveys.

When: Dolphin and whale research—year round, 2-month minimum commitment, for 2 days per week. REEF surveys—year round, for 1 to 2 hours per survey (free classroom training available—check upcoming dates at http://WildDolphin.org/events.html or contact; training materials sold at cost).

Who: Dolphin and whale research—individuals only, age 18 and over. Participants must be able to swim. REEF surveys—both individuals and groups welcome, all ages. Participants must be experienced snorkelers or divers.

Hardiness Level (5 being most difficult): Dolphin and whale research—2 to 4 (due to possible rough weather and exposure to elements while on board). REEF surveys—1 to 2.

Advance Notice: Dolphin and whale research—letters of interest accepted year round and held for when openings become available. REEF surveys—2 months.

Education: Onboard trainings are also available as part of a snorkel charters through Wild Side Specialty Tours.
(See http://sailhawaii.com/snorkel.html or contact WDF for details and fees.)

Donations: Monetary donations are appreciated and tax deductible. Contact regarding equipment and supplies needs. Ask about the Adopt-a-Dolphin program.

Contact: E-mail preferred. (Required content for letter of interest described at www.WildDolphin.org/volunteer.html.)
Tori Cullins
Wild Dolphin Foundation
87-1286 Farrington Highway
Wai'anae, HI 96792
(808) 306-3968
Tori@WildDolphin.org
www.WildDolphin.org

"A quote by Haunani Apoliona best exemplifies the vision and tireless efforts of [the] staff, value of both laypeople and grad student volunteers, work of researchers, contributions of marine and tour operators, and passion of those who love the sea, were born of the sea: *'Me na mea'oi loa mai nā wā mamua e holomua kākou i kēia au. . . .* Let us move forward into the future carrying with us the best from the past.'"
—Alexia Pihier
France

Acknowledgments

Mahalo to all the amazing people in this book, who've devoted their lives to giving back to our island environment. Your passion is contagious—it inspired me daily to complete this task of helping bring your efforts to greater light.

Thank you to the team at Island Heritage—Dale Madden, for supporting this vision; Micki Fletcher, for her creativity and patience, and her talented design team; Roxane Kozuma, Shara Enay, T. J. Steib, Mikayla Butchart, this was truly a collective effort; and especially to Jen Simpson, who has kept the tropics in her heart even from afar. *Mahalo* also to Cheryl Tsutsumi, whose experience provided me with signposts on the path, and to the Hāna Writing Circle, for their encouraging feedback.

Thank you to my mother, Sharon, for her infinitely constructive critiques and for always supporting my creative passions. And to my father, Harold, for giving me the practical tools to get from here to there. Lastly, to my husband, Rick, my confidant, my muse—*mahalo*.

Online Resources

When on the hunt for volunteer opportunities, look into these online resources. Some focus on geographical areas, such as particular islands, others on specific audiences, such as youth. As with the featured projects in this book, always confirm with the host organizations themselves that the information given online is up to date.

www.hawaiiycc.com
Hawaii Youth Conservation Corps is a free hands-on learning experience for youth with environmental groups on all main Hawaiian Islands.

www.hear.org/volunteer/maui
Hawaiian Ecosystems at Risk (HEAR) lists volunteer opportunities with environmental groups mainly on Maui.

www.kauaiyouthdirectory.com
Kaua'i Youth Directory matches youth with volunteer opportunities on the island. (Click on Volunteer Opportunities, then View by Category: Nature and Environment.)

www.malamahawaii.org
Mālama Hawai'i is a network of over seventy island organizations. (Click on Get Involved, then Volunteer Opportunities.)

www.MalamaKauai.org
Mālama Kaua'i provides a compendium of information about Kaua'i environmental groups. (Click on Get Involved.)

www.ponopacific.com/conservation3.html
Pono Pacific leads you to volunteer opportunities around the islands, plus provides information about critical environmental issues.

www.volunteerhawaii.org
Volunteer Hawaii is a matching resource provided by Aloha United Way for local volunteer opportunities. (You can narrow your search by Environmental, Animals and Environment, Cultural, and many other parameters.)

www.volunteerzone.org
Volunteer Zone is a volunteer clearinghouse for Hawai'i nonprofits. (Select Environment and your desired island from the home page.)

Index—General

'Ahahui Mālama i ka Lōkahi 120
Amy Greenwell Ethnobotanical Garden 21
Boo Boo Zoo, The 76
Clean Water Honolulu 122
Community Work Day Program 74
Earth Day 50
East Maui Animal Refuge 76
Farm Apprenticeships 116
FIN (Fish Identification Network) 111
Friends of Hakalau Forest National Wildlife Refuge 25
Friends of Haleakalā National Park 81
Friends of Hanauma Bay 124
Friends of Kamalani and Lydgate Park 53
Hakalau Forest National Wildlife Refuge 23
Haleakalā National Park 79
Hanalei Watershed Hui 56
Hawaii Audubon Society 126
Hawai'i Hawksbill Turtle Recovery Project 26
Hawai'i Nature Center 6
Hawai'i Nature Center—Maui 83
Hawai'i Nature Center—O'ahu 128
Hawai'i Service Trip Program 18
Hawai'i Volcanoes National Park 28
Hawai'i Wildlife Center 30
Hawai'i Wildlife Fund 7
Hawai'i Wildlife Fund—Hawai'i Island 33
Hawai'i Wildlife Fund—Maui 85
Hawaiian Islands Humpback Whale National Marine Sanctuary 2
Hawaiian Islands National Wildlife Refuges 5
Hawaiian Monk Seal Conservation Hui 58
Ho'oulu 'Āina 133
HOFA (Hawai'i Organic Farmers Association) 118
Honokōhau Valley Project 88
Honolulu Zoo Society 130

Hui o Laka 61
Ka Honua Momona 114
Kalihi Valley Nature Park 133
Kapahu Living Farm 93
Keālia Pond National Wildlife Refuge 90
Keep America Beautiful 8
Kīpahulu ʻOhana 93
Kohala Center, The 35
Kōkeʻe Natural History Museum 61
Kōkeʻe Resource Conservation Program 63
Lyon Arboretum 135
Mālaʻai 37
Malama Hanaula 97
Malama Honokowai 97
Mālama Na Honu 137
Mānoa Cliff Trail Project 140
Maui Coastal Land Trust 95
Maui Cultural Lands 97
Maui Nui Botanical Gardens 99
Maui Restoration Group 101
Mokupāpapa Discovery Center 39
Na Ala Hele 9
Na Ala Hele—Hawaiʻi Island 41
Na Ala Hele—Maui 103
Na Ala Hele—Oʻahu 142
Nani ʻO Waiʻanae 144
National Tropical Botanical Garden 10
Native Hawaiian Plant Society 105
Oʻahu Invasive Species Committee 145
Outdoor Circle, The 12
Pacific Whale Foundation 107
Project S.E.A.-Link 110
REEF (Reef Environmental Education Foundation) 109
Reef Check Hawaiʻi 14
ReefTeachers 35
Sierra Club—Hawaiʻi Chapter 17

Sierra Club—Kaua'i 66
Sierra Club—Moku Loa (Hawai'i Island) 43
Sierra Club—O'ahu 147
Surfrider Foundation 19
Surfrider Foundation—Kaua'i 68
Surfrider Foundation—Maui 112
Surfrider Foundation—O'ahu 149
Three Ring Ranch Exotic Animal Sanctuary 45
TREE (Tropical Reforestation and Ecosystems Education) Center Hawai'i . 48
Vegetation Program, Hawai'i Volcanoes National Park 28
Waihe'e Coastal Dunes and Wetlands Refuge 95
Waimea Middle School, Culinary Garden 37
Waipā Foundation 71
Wild Dolphin Foundation 151
WWOOF (Willing Workers on Organic Farms) 116

Index—Project Subject

Note that some organizations offer more than one type of project, so they may appear in multiple categories.

Beach/Site Cleanup and Beautification

Clean Water Honolulu 122
Community Work Day Program 74
Friends of Hanauma Bay 124
Friends of Kamalani and Lydgate Park 53
Hawai'i Wildlife Fund 7
Hawai'i Wildlife Fund—Hawai'i Island 33
Hawai'i Wildlife Fund—Maui 85
Hawaiian Monk Seal Conservation Hui 58
Keep America Beautiful 8
Nani 'O Wai'anae 144
Outdoor Circle, The 12
Pacific Whale Foundation 107
Sierra Club—Hawai'i Chapter 17

Sierra Club—Kaua'i 66
Surfrider Foundation 19
Surfrider Foundation—Kaua'i 68
Surfrider Foundation—Maui 112
Surfrider Foundation—O'ahu 149

Arboretum/Farm/Garden

Amy Greenwell Ethnobotanical Garden 21
HOFA (Hawai'i Organic Farmers Association) 118
Lyon Arboretum 135
Māla'ai 37
Maui Nui Botanical Gardens 99
National Tropical Botanical Garden 10
Waimea Middle School, Culinary Garden 37
WWOOF (Willing Workers on Organic Farms) 116

Habitat Restoration (Forests, Watersheds, Wetlands)/Trails

'Ahahui Mālama i ka Lōkahi 120
Friends of Hakalau Forest National Wildlife Refuge 25
Friends of Haleakalā National Park 81
Hakalau Forest National Wildlife Refuge 23
Haleakalā National Park 79
Hanalei Watershed Hui 56
Hawaii Audubon Society 126
Hawai'i Nature Center 6
Hawai'i Nature Center—Maui 83
Hawai'i Nature Center—O'ahu 128
Hawai'i Service Trip Program 18
Hawai'i Volcanoes National Park 28
Hawaiian Islands National Wildlife Refuges 5
Ho'oulu 'Āina 133
Hui o Laka 61
Kalihi Valley Nature Park 133
Keālia Pond National Wildlife Refuge 90
Kōke'e Natural History Museum 61
Kōke'e Resource Conservation Program 63

Malama Hanaula 97
Malama Honokowai 97
Mānoa Cliff Trail Project 140
Maui Coastal Land Trust 95
Maui Cultural Lands 97
Maui Restoration Group 101
Na Ala Hele 9
Na Ala Hele—Hawai'i Island 41
Na Ala Hele—Maui 103
Na Ala Hele—O'ahu 142
Native Hawaiian Plant Society 105
O'ahu Invasive Species Committee 145
Pacific Whale Foundation 107
Sierra Club—Hawai'i Chapter 17
Sierra Club—Moku Loa (Hawai'i Island) 43
Sierra Club—O'ahu 147
TREE (Tropical Reforestation and Ecosystems Education) Center Hawai'i . 48
Vegetation Program, Hawai'i Volcanoes National Park 28
Waihe'e Coastal Dunes and Wetlands Refuge 95
Waipā Foundation 71

Hawaiian Culture (Ethnobotany, Fishponds, Taro Patches)

'Ahahui Mālama i ka Lōkahi 120
Honokōhau Valley Project 88
Ka Honua Momona 114
Kapahu Living Farm 93
Kīpahulu 'Ohana 93
Malama Hanaula 97
Malama Honokowai 97
Maui Cultural Lands 97
Maui Nui Botanical Gardens 99
Waipā Foundation 71

Marine Environment (Ocean, Reefs)

FIN (Fish Identification Network) 111
Hawai‘i Wildlife Fund 7
Hawai‘i Wildlife Fund—Maui 85
Kohala Center, The 35
Mokupāpapa Discovery Center 39
Project S.E.A.-Link 110
REEF (Reef Environmental Education Foundation) 109
Reef Check Hawai‘i 14
ReefTeachers 35
Surfrider Foundation 19
Surfrider Foundation—Kaua‘i 68
Surfrider Foundation—O‘ahu 149

Marine Life (Dolphins, Fish, Seals, Turtles, Whales)

FIN (Fish Identification Network) 111
Hawai‘i Hawksbill Turtle Recovery Project 26
Hawai‘i Wildlife Fund 7
Hawai‘i Wildlife Fund—Maui 85
Hawaiian Islands Humpback Whale National Marine Sanctuary 2
Hawaiian Monk Seal Conservation Hui 58
Keālia Pond National Wildlife Refuge 90
Mālama Na Honu 137
Mokupāpapa Discovery Center 39
Pacific Whale Foundation 107
Project S.E.A.-Link 110
REEF (Reef Environmental Education Foundation) 109
Reef Check Hawai‘i 14
Wild Dolphin Foundation 151

Wildlife Rehabilitation/Zoo

Boo Boo Zoo, The 76
East Maui Animal Refuge 76
Hawai‘i Wildlife Center 30
Honolulu Zoo Society 130
Three Ring Ranch Exotic Animal Sanctuary 45

Index—Time Commitment

Note that some organizations offer more than one type of project, so they may appear in multiple categories.

A Few Hours to One Day

'Ahahui Mālama i ka Lōkahi 120
Amy Greenwell Ethnobotanical Garden 21
Clean Water Honolulu 122
Community Work Day Program 74
FIN (Fish Identification Network) 111
Friends of Hakalau Forest National Wildlife Refuge 25
Friends of Hanauma Bay 124
Friends of Kamalani and Lydgate Park 53
Hanalei Watershed Hui 56
Hawaii Audubon Society 126
Hawai'i Nature Center—O'ahu 128
Hawai'i Volcanoes National Park 28
Hawai'i Wildlife Fund—Hawai'i Island 33
Hawaiian Islands Humpback Whale National Marine Sanctuary 2
Ho'oulu 'Āina 133
Honokōhau Valley Project 88
Honolulu Zoo Society 130
Hui o Laka 61
Ka Honua Momona 114
Kalihi Valley Nature Park 133
Kapahu Living Farm 93
Keālia Pond National Wildlife Refuge 90
Kīpahulu 'Ohana 93
Kōke'e Natural History Museum 61
Māla'ai 37
Malama Hanaula 97
Malama Honokowai 97
Mānoa Cliff Trail Project 140
Maui Coastal Land Trust 95
Maui Cultural Lands 97

Maui Nui Botanical Gardens 99
Maui Restoration Group 101
Na Ala Hele—Oʻahu 142
Nani ʻO Waiʻanae 144
Native Hawaiian Plant Society 105
Oʻahu Invasive Species Committee 145
Pacific Whale Foundation 107
Project S.E.A.-Link 110
REEF (Reef Environmental Education Foundation) 109
Reef Check Hawaiʻi 14
Sierra Club—Kauaʻi 66
Surfrider Foundation—Kauaʻi 68
Surfrider Foundation—Maui 112
Surfrider Foundation—Oʻahu 149
Three Ring Ranch Exotic Animal Sanctuary 45
TREE (Tropical Reforestation and Ecosystems Education) Center Hawaiʻi . 48
Vegetation Program, Hawaiʻi Volcanoes National Park 28
Waiheʻe Coastal Dunes and Wetlands Refuge 95
Waimea Middle School, Culinary Garden 37
Waipā Foundation 71
Wild Dolphin Foundation 151

A Few Days to a Few Weeks

Friends of Haleakalā National Park 81
Hakalau Forest National Wildlife Refuge 23
Hawaiʻi Service Trip Program 18
Kapahu Living Farm 93
Kīpahulu ʻOhana 93
Lyon Arboretum 135

At Least One Month

Lyon Arboretum 135
Mokupāpapa Discovery Center 39

At Least Two Months

Hawai'i Hawksbill Turtle Recovery Project 26
Keālia Pond National Wildlife Refuge 90
Māla'ai 37
Mālama Na Honu 137
Surfrider Foundation—O'ahu 149
Three Ring Ranch Exotic Animal Sanctuary 45
Waimea Middle School, Culinary Garden 37
Wild Dolphin Foundation 151

At Least Three Months

Hawai'i Volcanoes National Park 28
Honolulu Zoo Society 130
Vegetation Program, Hawai'i Volcanoes National Park 28

Flexible Time Commitment

Boo Boo Zoo, The 76
East Maui Animal Refuge 76
Haleakalā National Park 79
Hawai'i Nature Center—Maui 83
Hawai'i Wildlife Center 30
Hawai'i Wildlife Fund—Maui 85
Hawaiian Monk Seal Conservation Hui 58
HOFA (Hawai'i Organic Farmers Association) 118
Kohala Center, The 35
Kōke'e Resource Conservation Program 63
Na Ala Hele—Hawai'i Island 41
Na Ala Hele—Maui 103
National Tropical Botanical Garden 10
Outdoor Circle, The 12
ReefTeachers 35
Sierra Club—Moku Loa (Hawai'i Island) 43
Sierra Club—O'ahu 147
Waipā Foundation 71
WWOOF (Willing Workers on Organic Farms) 116